HONEYMOON

Dorothy May Mercer

Dorothy May Mercer

Kelly & Tom

i

ISBN 13: 978-1-62329-078-8
ISBN 10: 1-62329-078-3

© Copyright 2018, 2019 by Mercer Publications & Ministries, Inc.
Library of Congress Control Number: 2018914178

Publisher: Mercer Publications & Ministries, Inc.
Stanwood, Michigan USA 49346-9644

This is a work of fiction and is entirely the imagination of the author. Any resemblance to persons living or dead is purely coincidental.

"Hail to the Chief" started playing. A cry went up from the people closest to the stage entrance. Wild cheering swelled as the crowd realized who had entered. Cameras swung onto the president and his wife, First Lady Janette Bigelow, as they entered hand in hand. Gerard smiled broadly while waving at the crowd, strolling across the stage, and pointing from time to time.

"USA! USA! USA!" screamed the crowd.

The sound switched instantly as President Gerard Bigelow stepped to the mike. "Thank you, thank you." The applause died down. "Hello, Kansas!" he exclaimed, followed by another full minute of cheering. He waved both arms, grinned, paced, and nodded.

The audience knew when to cheer, and the president was an expert at working the crowd. "I'm pleased to announce the largest military budget in history, seven hundred billion dollars. We have the greatest military force in the history of the planet." The audience roared its approval.

Meanwhile, across the oceans, on the other side of the planet, Dear Leader (DL) and General Lee listened to Bigelow's speech. "Not for long," DL grimaced. "Soon the tiny little mouse is going to take down that big, old, fat elephant."

General Lee chuckled and nodded agreement.

Table of Contents

Prelude

Panic Call

"Mike? Mike, it's me. Mike, something's going on here," said a soft feminine voice, barely above a whisper.

"Who's calling?" Mike demanded.

"It's me. Mike, I'm scared."

"Speak up. I can barely hear you."

"Mike, I can't. They'll hear me," she said, getting closer to the phone. Her voice sounded young and very frightened.

Crusading United States Senator Michael McBride didn't have time to fool around with prank calls. Rarely did one ever slip through his outer office to Cynthia, his efficient personal secretary, much less to his desk. He was about to hang up.

"Mike! Please! Don't hang up. It's me," she repeated, panic causing her voice to rise.

Suddenly realization clicked in Mike's brain. Only family members and a few of his closest friends could get through directly. This had to be one of his two sisters.

"Sis," he said, guessing correctly. "What's going on?"

"They saw me," she said, panic rising. "They may even know I'm calling you." She nearly sobbed.

"Who is' they'?"

"Never mind. Listen, I can't say any more over the phone. Mike, we've got to stop them. It's terrible. I saw what they are planning to do. It's just ... it's just ... Oh my God, Mike, unbelievably horrible."

Mike was thinking fast. It had to be his younger sister. "Where are you, Kelly?"

"I'm across from their lab."

"Aren't you with Tom?"

Kelly and Carson City Police Sgt. Tom Turbulo were newlyweds. Senator Mike McBride had returned from their wedding in New Mexico, just two weeks ago. They were still on their honeymoon as far as Mike knew. He had no idea where they had gone. Some island somewhere, he thought, but he wasn't sure.

"Mike, I've got to hang up," she whispered hurriedly. The line went dead.

Dorothy May Mercer

E M P Honeymoon

Kelly & Tom

Chapter 1

Barefoot Sands Resort and Spa, Shoreline Road, Honduras

Back in their rented villa, Kelly hastily locked the door, throwing the deadbolt closed, adding the chain. She leaned back against the door, chest heaving as she recovered her breath. In a moment, she hurried about their secluded oceanfront cottage to check all the windows, making certain they were securely locked.

If only Tom would hurry, she thought. *God, I hope and pray he's safe.*

Kelly moved into their bedroom and threw her phone on the bed. Staying out of sight as she pulled the drapes closed, she then drew aside one corner so she could peek out at the beach area in front of their villa.

Craning her neck first one way and then another, she watched for several minutes, not knowing what to expect. *What if they followed me? Who can I trust?* Alone, Kelly knew she could not defend herself against more than one man. She was a sitting duck, lucky to have escaped, but it wouldn't be so easy next time.

Kelly had spent hours and weeks choosing this place for their honeymoon. It was on a Caribbean island, with

3

gorgeous beaches, a dive center, delicious meals, and their own private cottage right on the beach. Unlike so many private island resorts, this price was within their budget. Kelly could not believe what similar places cost. She thought the posted prices were terrible for a week, until she learned that the price was for one night. Kelly had almost given up when she found this affordable place. What Kelly did not know was the reason it was cheap.

Normally, Tom and Kelly kept the doors and windows of their villa wide open so they could enjoy the ocean breezes and the hypnotic sound of the waves lapping the sandy shore. Alone now, Kelly felt trapped, doors and windows locked tight, just waiting for God-only-knew what.

Her new husband, Carson City Police Department (CCPD) cop, Sgt. Tom Turbulo was out with the dive boat, exploring the world-class coral reefs, probably having the time of his life.

Earlier, Kelly had opted to take an excursion bus around the island to stroll through some of the shops and local venues. The tour director carried on a running commentary in Spanish and broken English. Kelly had a passable knowledge of Spanish so long as the speaker talked slowly and distinctly. She had heard enough to gather that they were stopping for twenty minutes, with the warning to stay close-by on the main street and be back here on time.

She should not have wandered so far away from the tour bus, onto a side street. Curious by nature, she was intrigue by the window display. It was a deceptively normal-appearing, tiny gift-shop with an unusual storefront. The shop called La Pocilga—The Pigsty—sported a tall, artfully decorated fake-front and a window display of dozens of toy pigs. She had no idea it concealed

4

a strange high-tech lab behind which an expensive vehicle was parked next to a shipping container.

None of the locals really knew what was going on behind the shop, nor did anyone care, so long as the necessary palms were greased, and high rents paid.

Kelly had stepped into the shop, thinking she might find the perfect piggy gifts to take home to her nephews and nieces. She thought she might even find something nice for her mom, Grace McBride, and her new mother-in-law, "Mom" Turbulo. Kelly had to start thinking of her as Mom, and not Mrs. Turbulo. *I'm Mrs. Turbulo, now,* Kelly reminded herself with a grin. It had taken her long enough to "land" that handsome cop for her husband, but the wait was worth it. She was crazy in love, never happier.

Soon Kelly was so engrossed in selecting gifts she did not notice she was alone or realize the tour bus had moved on to the next stop. Arms laden, she looked around for the clerk. Seeing no one, she pulled aside a curtain and walked down a hall, first tapping on doors, and then peeking into a small restroom and closets filled with goods and various cleaning tools. Finally, she moved up to a heavy door marked *Danger Keep Out, Employees Only*, in Spanish and English.

Should I look out here? She wondered, ignoring the warning. *I'm sure the clerk was here just a minute ago. He must have taken a break outside for a smoke.*

Cautiously, Kelly elbowed the door, thinking to call, "Hello?" Instead, at the first sight, her jaw dropped a mile. Eyes popping, she breathed a silent "Ah-oh, my!"

Quietly withdrawing, Kelly hastened to drop her packages on the counter and tiptoe back to the forbidden door. Curiosity overcoming caution, this time she opened the door just a tiny crack, to remain hidden from sight.

Seeing and hearing no-one, she widened the opening, ever so slowly, and slipped inside.

In all her university days, she had never seen a lab like this. Banks of quietly humming computers, huge monitors scrolling cryptic formulas, rows of mysterious machines. Overhead were several enormous interactive screens displaying maps of the United States and Canada, two of them showing thousands of interconnecting gridlines and blinking colored lights.

Kelly knew she had seen displays like this somewhere before. Her brain raced, trying to make the connection. *What is this? I know I've seen a picture like this somewhere.* Kelly bit her lip trying to bring it up. Then she remembered, *Oh yeah, in one of my classes on homeland security. Something about our vulnerability... It was Professor Kin...Kinno...Kinner... uh... what was his name? Peter something ...Oh never mind it doesn't matter. It had to do with the power grid. That must be it. These two are the North American power grid. But, what are they doing with it here? Why here, out in the middle of nowhere?*

Kelly's gut clenched. *This can't be good.* She patted her pockets, about to reach for her phone when a rough hand covered her mouth. A strong arm crushed her from behind, pulling her off-balance, and dragging her across the floor toward the back door.

Kelly opened her mouth to scream. A wrist crammed deeper inside her mouth. Kelly tried to bite. The hand and wrist jammed her throat harder making her gag.

Her attacker had to let one hand go momentarily to open the exit door.

Summoning her fading strength, Kelly butted with her head and managed a sharp connection with one elbow.

"Oof," the man's diaphragm caved.

In that split-second Kelly's feet connected, and she took off, heavy footsteps in hot pursuit. The man grabbed her backpack. Kelly slipped out of it and ran.

Outside, Kelly darted across traffic, dodging cars and trucks. She faded behind a van and quickly ducked into a storefront where she could observe her pursuer.

Watching in-between breaks in the traffic, she could see a burly man standing in front of the gift shop, hands on hips, angrily whipping right to left, scrutinizing every passerby and barking orders into a communicator.

Staying out of sight, Kelly had grabbed her phone and pressed Mike's call number. No sooner had she started to explain when she noticed a half-dozen armed men joining her pursuer. Within seconds they had fanned out searching door to door.

Kelly jammed the phone into her money-belt and darted toward the back entrance, moments ahead of the search party. All she could do was run for her life. Luckily she still had her money-belt where she kept her passport, phone and important papers. Her tour bus was nowhere in sight. Ahead was an entrance to a hotel. A couple of taxicabs idled out front. Approaching the hotel, Kelly must have looked frazzled. She tried to slow her pace and appear normal. Opening the door of the first cab in line, she blew out a breath and clambered inside.

"Take me to the fishing pier," she said, first in English, then in Spanish.

"Yes, Miss, sí señorita," the cab driver replied, as he put his taxi in gear and drove forward.

Kelly leaned back, sighed heavily and began to quiver.

It would be a ten-minute ride to the fishing pier if there weren't so many people on bikes and walking in the street. Kelly's driver alternated between dodging potholes and people while blowing his horn and cursing.

Kelly had time to calm down and think. At first her plan had been to change her instructions once they were underway. But now she realized it was good that the driver did not know her address. No doubt her pursuers would question him later. Kelly would get off at the pier and catch a city bus to her resort.

Senate Office Building

Senator Mike McBride stared at the phone in his hand. What had just happened? Acting quickly, he tapped the screen to recall the previous incoming number. "Trace call," he commanded. Kelly's number popped up. "Location," he said.

His phone reported, "The latitude of West Bay, Honduras is 16.276478, and the longitude is -86.596840."

"Show on map," said Mike, struggling to remain calm and think straight. In his lifetime as a cop, now a senator, Mike had faced many crises having to control himself in any possible emergency, but this was his baby sister. He fought to slow his heartbeat, be rational and use reason.

Okay, she's in Honduras, he thought. *She's with Tom and Tom's a smart cop. For some reason, she thought she had to call me. Either she wasn't with Tom, or it had to be more than Tom can handle. Something to do with national security, maybe? No, of course not. Why would I think that?*

Mike picked up his phone and recorded a text message to Kelly, "Where are you? Where is Tom? What's wrong?" and clicked Send.

Then he sent another one to Tom, "Are you with Kelly?"

Setting his cell phone aside, but within reach, Mike used the encrypted desk phone to call his friend at the CIA.

"Rench here," said Assistant Director Joseph Rench.

"Jo, this is Mike McBride. Got a minute?"

"Mike! How are you?"

"Not so good at the moment. I just had a strange call from my sister, Kelly."

"Not your baby sister?"

"Not a baby anymore, Jo. She's all grown up and married now."

"Oh, darn it. Too late, I missed my chance," said Jo with a chuckle.

"Jo, what can you tell me about West Bay, Honduras? Is anything going on down there?"

"Could be. Or not. Hard to tell. Honduras is in bad shape, right now," said Jo.

"Come on, Jo. I'm serious."

"What's this got to do with Kelly, anyway?"

"I think she's down there on her honeymoon."

"Well, there are some fairly safe island resorts there, catering to tourists. But parts of the mainland are considered dangerous for Americans. So, Mike, just relax. Let nature take its course."

"Jo, please. Who have you got down there?"

"Now, Senator, you know better than to ask that."

Mike sighed in exasperation. "You aren't helping."

"Sorry, pal. Try a different question."

"Okay, let me think. Uh, has the satellite been watching over activities in Central America?"

"Of course."

"Honduras?"

"Probably."

"And that would be because of suspicious activity going on. No, correct that…unusual activity."

"Pretty much, always. Why?"

"Jo, my sister has stumbled upon something. I'm sure of it."

"Maybe I need to talk to her."

"Why do you say that, Jo?"

"That's all I can say, Mike."

"I see."

Jo's voice sounded concerned. He said more with his tone of voice than with his actual words. "Tell her to enjoy her honeymoon, stay close to her hotel and mind her own business. The tourist places are safe, but it's a jungle out there."

"What is that supposed to mean?"

"Gotta go. Just take care of her, Mike."

Mike's cell phone chirped with an incoming message.

It was from Tom. Mike opened the message, "On dive boat. What's wrong? Tom."

"Wherz Kelly?" Mike tapped.

"Gone 3-hour tour," Tom responded.

"Hold on. I'll call u," Mike wrote. He hit the phone icon and Tom's number.

Tom answered almost before it rang. "What the hell is going on, McBride?"

"Heh, bro, your wife called me in a panic. Something happened to her in town. She sounded terrified and breathless like she had been running. She's not answering my text, but maybe she'll pick up if you call her."

"Jesus!" Tom exclaimed. He started punching buttons in panic mode, unable to hit the right button.

The first mate came up beside him. "Can I help you, sir?"

"I c-can't …" Tom swiped and punched several more spots on the display. "M–my wife!" His voice shook. "I … I …"

"Here, let me do it," suggested the mate with a smile. "We'll have your wife in a jiffy." Quickly he cleared the phone with one swipe, clicked Contacts, tapped the right access code and the first number on the list. "It's ringing," he smiled gently and laid a hand on Tom's arm. "Hang in there, buddy. Just a sec."

"Hello, Tom? Oh thank God, it's you," a feminine voice answered.

"Here you go, sir," said the mate, handing the phone to Tom. "Say hello," he said and turned away.

"Kelly?"

"Yeah, it's me."

"You all right?"

"Uh…"

"Mike called."

"Sorry. Didn't mean to upset anyone."

"Well, what's going on?"

"Um…"

"Where the hell are you?"

"I'm back at our place," she said, trying to sound normal. "When are you coming in?"

Tom looked at his watch. "Probably in a couple hours. Why?"

Kelly's mind was finally working. She realized she couldn't say anything to Tom over the phone, lest he tip off the crew members. At this point, probably the bad guys were looking all over for some unknown woman, asking questions. All they had was a vague description. She needed to be calm and stay out of sight.

"Oh nothing," she lied. "I'm fine and missing you already, sweetheart."

"Miss you, too."

"Well, I'll sign off. I need to take a shower and get into something comfortable. You have fun and catch a big one, okay? Maybe I'll take a nap and rest up for our evening."

"We're not fishing, honey. You sure you're all right?"

"Absolutely. Couldn't be happier, unless you were here, of course."

"Okay, then," said Tom, doubtfully. "If you're okay without me, I'll see you in a couple hours."

"I'm sure. Bye-bye, sweetie."

Tom shoved the phone into his pocket. Turning toward the mate, he shrugged and grinned. "New wife," he explained, sheepishly.

The mate nodded in sympathy and turned back to helping the other divers with adjusting their masks and oxygen lines.

Chapter 2

State University, Kinney Memorial Hall

Dr. Peter Kinney turned from writing on the white board. Addressing his students, he announced, "That will be all for today, class. I'm off to New York for the weekend. You may want to catch my interview Sunday night at 10 PM Eastern time. That is, if you have nothing better to do," he added with a wry smile.

The students laughed. A few actually applauded.

Peter turned to his desk to gather up his papers, disregarding the hum of conversation while his packed lecture hall emptied of students. His was one of the popular classes on campus and he was one of the most well-liked professors.

Self-deprecating, he liked to joke about how he inherited his job because his great-grandfather endowed Kinney Hall and the Kinney teaching chair he occupied in advanced nuclear physics. The truth was his reputation was golden as one of the world's top experts in his field, and yet he managed to inject humor into his talks, while making sense of an extremely difficult and complex subject.

Peter knew and taught the incredible benefits to humanity of nuclear energy, but at the same time his private passion was with the destructive power it possessed, in the wrong hands. He viewed with growing alarm the threat posed by certain third-world dictators to the United States and the free world.

Peter hated to leave his mountainside home. It was his paradise, where he retreated from the world with his dogs, books and his devoted wife of thirty years. But, this issue was too important. His scholarly papers and articles in obscure journals simply weren't enough to get through to the political powerbrokers in Washington D.C. He had to do more to spread the alarm and ignite a fire in the Defense Department. That is what led him to accept the invitation for an hour-long interview on the Haven-Harbinger TV show, broadcasting from New York City.

Channel Nine, New York

Henry Haven-Harbinger's reputation was as an honest reporter and journalist, reminiscent of the old-fashioned journalists such as Barbara Walters. His TV shows focused on in-depth interviews with experts in a different scholarly field each week. Haven-Harbinger was totally prepared for each interview, and yet he never spoke in "sound-bites." Instead, he had a way of bringing out the best in his guests, neither arguing nor challenging them, and never ridiculing or insulting in the manner of most present-day TV talk-show hosts. Henry did not need to use any tricks to generate an audience. His respectful manner and reputation as a fair interviewer were such that he was able to gain access to the very best and some of the most difficult scholars to reach. Surprisingly, for such a laid-back style, his show was the top-rated cable show in its time slot.

Dr. Kinney listened, off-camera, as Henry Haven-Harbinger introduced him. "My guest for this evening is little known outside his field. But, within his field and among his peers there is no one more highly regarded. We understand that Dr. Peter Kinney never grants interviews, so we are very lucky to have him here tonight, in person.

"Dr. Kinney is a tenured Professor of Nuclear Physics at State University in Utah. He holds a doctorate from the Massachusetts Institute of Technology and a Law degree from Harvard. He also holds a doctorate in Astronomy from the University of Alberta in Calgary, Alberta Canada. He has received the prestigious Nobel Prize in Physics for his ground-breaking work and discoveries in the field."

The camera swung to a view of Peter as he walked onto the set.

"Welcome to our show, Dr. Kinney," said Henry as he rose to shake hands. "Please be seated."

"Thank you for having me," said Peter as he shook hands and took a seat in a comfortable upholstered chair opposite the host. The set was designed to appear like two friends seated in a living room, relaxing in front of a fireplace.

"Our audience has been told of your amazing accomplishments, Dr. Kinney, but what they don't know is your other interest, your hobby if you will. There has to be a compelling reason for you to leave your mountain paradise and travel all the way to the big noisy city."

Peter laughed. "You are quite right, Mr. Harbinger."

"None of this 'Mister' business. Call me Henry."

"And please call me Peter. Dr. Kinney was my grandfather," he grinned.

"And so, Peter, let's get right to it. Suppose you tell us what brings you here."

"You're right, Henry. I want the American people to know what can happen if the power goes out."

"But, don't we already know? Everything goes dark, right?"

"Oh, yes, we've seen minor events. For example when New York and large parts of the eastern seaboard went dark for three days back in '02."

15

"Well that wasn't so minor to those people stuck in elevators, or for patients in surgical operating rooms."

"I see what you're saying," Peter agreed, "but hospitals are prepared for this. In most cases there was nothing more than a blink as the service switched to backup generators."

"I see."

"The people in elevators were rescued with no injuries. They had something exciting to write home about. And don't forget the side benefit of the population boom which occurred nine months later."

"True," Henry chuckled.

"What people don't realize is that many of our North-American power grids are all interconnected. At that time the Eastern seaboard was rescued by power from Canada. That was a good thing.

"But what worries me is how vulnerable we are. In the last ten or twenty years both Canada and the United States have moved into technology at warp speed. Think about it—the changes which have come about, in just our lifetimes. It isn't just cars, airplanes, banks, and the internet. More and more every-day systems are being controlled and operated by computers, as well. And remember this," he raised one finger for emphasis, "this is the key to our vulnerability: computers are run on electricity.

"Can kids do simple math? Can clerks make change anymore, or do they need a machine to tell them how many bills and coins to count out?"

"You're right," said Henry, congenially. "Sometimes the coins come out automatically, too."

"Right, and powered by electricity."

"That's true."

"Now some large stores have completely done away with clerks. You scan your own purchases, tap a screen, swipe a debit card, or insert a bill. The machine grabs your bill and spits out your change. So in view of all that, what happens if the power goes out?"

"I suppose the generator kicks in."

"Right, but sooner or later the generator runs out of gas."

"And so do the customers," Henry added, pausing thoughtfully.

"My daughter teaches high school," Henry offered. "At a family dinner, Sunday, she talked about the training she was taking to learn how to switch her classroom over to a new computerized smart phone system for teaching her students. It seems all the kids are wired in, nowadays. I don't even understand it."

"Right, and I was just into my doctor's office for my annual checkup. I've been going to the same place for twenty years. No longer do they have those long banks of paper files lined up on the wall. Nor do they need three medical secretaries to handle it. I had to repeat all my information and meds, so one young girl could input it into their new software system. She ran my Social Security card and my driver's license through a special little machine, so it could upload into a computer storage system in some distant city. And my P.A. spent all our time together typing stuff onto her laptop computer. I doubt if they ever write out a prescription on the old-style paper pads anymore.

"I tried to hand her a paper print-out of the latest test I had run at the hospital. 'Never mind,' she said. 'I have that all right here on my computer.' I'm thinking 'what happened to privacy'."

"You're right," said Henry. "Even the North Koreans can read about my hemorrhoids."

"Exactly," Peter laughed. "But, my concern is more than a loss of privacy. It's just this: what happens when the power goes out?"

"Surely the United States government has a backup plan in case of emergency. Don't they? Isn't that what FEMA does? You know, help out in a crisis?"

"You mean the Federal Emergency Management Agency?"

"Yes."

"They are a pretty big outfit. I understand they have an annual budget of more than thirteen billion dollars."

"Wow." Henry gave a low whistle.

"That's not all. States and some large municipalities have their Emergency Agencies. And don't forget the O.E.M., the Office of Emergency Management. This is a separate agency charged with planning."

"I get it. What you're saying is the bureaucracy is too huge to be efficient?"

"No, I'm not saying that at all. They do a pretty good job of coordinating in case of emergency. Oh sure, there are always critics, but, all in all, they work together well, thanks to interconnecting computer systems, run by electricity."

"Computer systems, again."

"Exactly. All I'm asking is 'what if the power goes out'."

"But, doesn't that happen? I mean look at Hurricane Sandy or Hurricane Katrina. Didn't the power go out?"

"Sure. Of course it did. Power lines were down all over. But, I'm not speaking about mere power lines. Those can be rebuilt in a few days, or weeks at the most. In neither of those cases did the main power generator go out."

"Well, no I guess not," Henry agreed.

"The North American power grid is divided into a half dozen or so sections. Each section is handled by enormous super-generators, which control and disburse the electricity to sub-stations, which, in turn, disburse it to smaller transformers which send it out over lines into our houses, factories and office places."

"I've seen those small gray boxes when I go out for a walk."

"Right. And you may have seen a power sub-station when you went out for a drive in the country."

"Yes, I have."

"Was it surrounded by a chain-link fence, with a 'Danger Keep-Out' sign on a locked gate?"

"Yes, I believe it was."

"Have you ever seen an armed guard patrolling the fence?"

"Well, no. I haven't." said Henry. "You're kidding, of course."

"Imagine the fireworks one hand grenade could cause," said Peter, tongue-in-cheek.

"Now you're scaring me."

"I'm worried about a much bigger fireworks, man-made or natural, knocking out a super-generator. If we lost one of those it takes months to order a new one. There is only one country manufacturing them."

"You mention man-made and natural, meaning terrorists might try, I suppose. But what possible natural force could knock out a sub-station, much less a super-generator? Hurricanes, floods, tornados?"

"Most likely none of those could hurt a sub-station. But one natural force could do it, easily. It's the biggest, most natural source of power that we have. It's right there before our eyes every day. It's just there, so dependable we take it for granted. We don't notice it unless we are so stupid as to look directly at it."

"The sun?"

"Yes, the sun can wipe out our entire power grid and all our systems with one medium sized solar flare."

"Come on, Peter. That can't happen."

"But it already has."

"I didn't know that."

"Of course, there are solar flares quite often. Minor ones, so to speak although even a minor one is huge, spewing trillions of tons of energy thousands of miles. For the most part we are protected by our atmosphere, although there can be disruptions in communication and a beautiful display of aurora, called Northern Lights. After all, our sun is a star, one gigantic atomic explosion going on continuously.

"But what I'm referring to is a solar super-flare of the kind that only occurs every few centuries. Think of a volcano such as Yellowstone, spitting steam and rumbling away for centuries until, one day, it explodes."

Henry quietly took this all in. "Well, surely the scientists know about this," he said.

"There are studies," answered Peter. "The United States and others have observatories and deep space probes studying the stars in an effort to learn as much as possible about flares and super-flares. One such space probe studied 83,000 stars over a ten-year period, collecting data on types and behavior of stars, their flares and super-flares."

"What is the difference?"

"The main difference is that a super-flare is about 10,000 times more powerful than a flare."

"Have any hit the earth?"

"The first *recorded* solar flare occurred in 1854, the so-called Carrington flare, named after the man who discovered it. It was a mere one ten-thousandth as strong as the strongest outer-space-flare recorded by our space probes, but still it knocked out the telegraph, which was in its infancy back then. Also, the aurora was seen over most of the northern hemisphere."

"Of course, this was the horse and buggy era. Mail was delivered by pony express. Right?"

"Indeed it was," Peter chuckled. "Imagine the mess if such a flare occurred today."

"No more Netflix. How would our kids survive?"

"They might have to go back to textbooks and whiteboards."

"I shudder to think of it."

"Scientists estimate that if the Carrington solar flare occurred today it could damage or destroy our satellites. Persons on trans-polar flights would receive high doses of radiation, as would astronauts or the crew of the international space station. There would be significant damage to the ozone layer and damage to the biosphere which governs photosynthesis in the oceans. There could be a failure of the electricity distribution system with damage to transformers and switching equipment as well as loss of power to the cooling systems of spent fuel rods stored at nuclear power stations, loss of most radio communication because of increased ionization in the atmosphere.

"Even if that happened, in time we could recover and survive a solar flare of the size of the Carrington flare. What worries me is the devastation that would occur from a solar super-flare, one that is up to ten-thousand times greater."

"What might that be?" asked Henry.

"Pretty much the instant destruction of our civilization. For one thing, everything that is run by computer would melt and die. Cars, trains, trucks and buses would stop, airplanes fall from the sky, amusement parks collapse, elevators crash. Fires and explosions would break out. Imagine the Chicago fire everywhere and a long dark winter with no plants, animals or water."

"Would any life survive?"

"I suppose anything is possible, but for how long and for what purpose? Once the atmosphere is gone, the earth would go into a deep freeze, like the moon. If there was any water left, it would turn to ice."

"So how likely is it that our sun will ever have a super-flare?"

"Well, the space studies place suns into categories, according to various criteria, such as size, age, and spin velocity. The data shows that super-flares are more common in so-called younger, faster-spinning stars. Even so, it does happen in stars such as our sun, just not as often.

"Of course not all the sun's flares are directed our way. Scientists know that on July 23, 2012 the sun had an unusually large and strong so-called coronal mass ejection (CME) event which barely missed the earth. It erupted on the other side of the sun, as it rotated away from the earth. They say it missed the Earth with a margin of approximately nine days, as the Sun rotates around its own axis with a period of about 25 days. This eruption was on the order of the size of the Carrington flare.

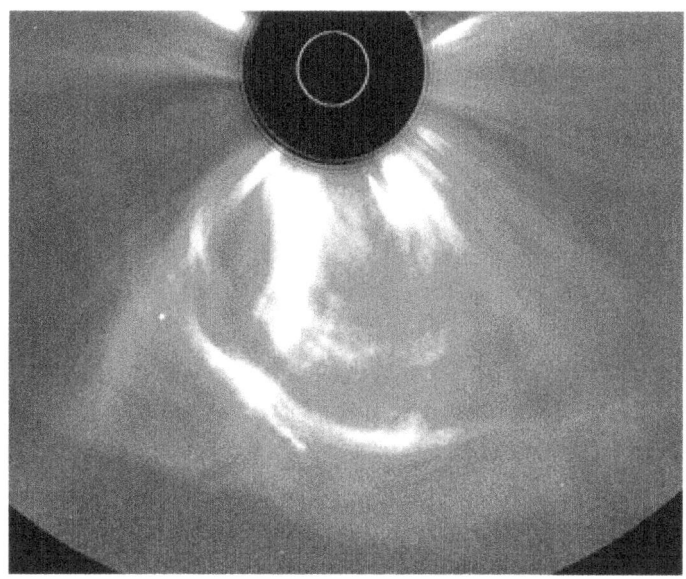

Photo of the sun's flare, July 23, 2012, by NASA/STEREO - http://newscenter.berkeley.edu/2014/03/18/fierce-solar-magnetic-storm-barely-missed-earth-in-2012/, Public Domain, https://commons.wikimedia.org/w/index.php?curid=37792812

"However historical evidence shows that a much larger solar super-flare may have hit the earth as late as 754 AD."

"How so?"

"There are ways of determining that by studying glacial ice from that period, and tree rings from spots all over the earth. Rather surprising supporting evidence is found in historical writings that describe strange visions in the sky around that time-period and further south than the aurora usually appears."

Henry spoke directly at the camera, "We have to take a break. We have been speaking with Nobel Laureate Dr. Peter Kinney. Thank you, Dr. Kinney for that chilling description of what can happen if our sun erupts in a super-solar flare. When we come back, I'll ask Dr. Kinney what he predicts the future might bring. Don't leave us. We'll be right back."

Barefoot Sands Resort

Relaxing in their villa, Kelly and Tom were watching a streaming re-broadcast of the show—one of Tom's favorites—on their computer. "What do you think of that?" asked Tom, as he fast-forwarded through the commercials.

"Very scary," she admitted, "but there is no sense worrying about something over which we have no control. Right?"

"Yeah, I agree. If a solar super-flare happens, we'll all be gone anyway. Might as well carry on and hope for the best."

"So, what do you say? Let's fix something to eat. I'm starved."

Always a cop, for some reason Tom's intuition was blinking yellow. "Can we watch the last half of the show, first? I'm curious about what this Kinney-guy says next."

"Okay, all right. Turn up the sound. I'll fix us a snack."

TV Studio

After the commercial break, Henry Haven-Harbinger spoke. "Welcome back, ladies and gentlemen. We are here with Professor Dr. Peter Kinney.

"During the break, Dr. Kinney told me about a monumental threat to our nation that is going almost totally ignored outside of scientific circles. Please share this with our viewers, Peter."

"Well, it has to do with the fact that our politicians and our military are still fighting with a cold war mentality. You know--more soldiers and better ships, airplanes and submarines."

"But, don't we have a missile defense system?"

"Yes, we do, but it was built with the expectation of an attack from the old Soviet Union."

"Well, Russia still has its weapons."

"Yes, and so our missile defense is aimed at the north, leaving our southern flank wide open. Not only that, we have no defense against an attack from on high."

"Space?"

"Exactly."

"But, what harm can a tiny satellite do?"

"We have no defense against a manmade E.M.P."

"E.M.P.?"

"It stands for Electro-Magnetic Pulse. It's a kind of twenty-first century warfare. Our enemies no longer need to crash airplanes into buildings or drop bombs on our cities. They merely need to figure out ways to sabotage the internet and wipe out our computer-systems. An electro-magnetic pulse on the order of..., uh... let's say the power of, a strong solar flare could turn us back to the dark ages.

"At that point, if there was anything left, an invading army could simply move in and take over before we knew what happened. This army would have to come equipped with its own systems and be able to set up shop."

"This sounds like science fiction."

"Don't think it can't happen. We believe countries can design newer, smaller tactical uses of light-weight nuclear weapons that can be mounted on intermediate range ballistic missiles or mini-satellites and exploded remotely."

"But, how would this work?"

"Theoretically, a small atomic explosion, two or three hundred miles up, would not even be heard or seen on earth because it would occur in a vacuum."

"Something like the old question," said Henry. "If a tree fell in a forest and there was no one to hear it ..."

"Does it make any sound?" With a laugh, Peter completed the conundrum. "I think you are getting the picture."

"So, all our enemies have to do...," said Henry

"Or friends," reminded Peter.

"All our enemies or friends have to do is figure out ... uh ... what?"

"They have to figure out how to limit the effect of an EMP (Electro-Magnetic-Pulse) to a target, let's say North America."

"And Europe?"

"Well, maybe. I suppose that depends on who is setting it off. It will be a country on the other side of the world, I suppose, or one that is so backward that it is not dependent on technology."

"So, how will we know? I mean, there are hundreds of satellites going around. How can we tell whether the satellite is friendly or not? We can't just shoot satellites out of the sky, willy-nilly."

"We won't know, will we?"

"So what can we do?"

"Well, I'm not privy to any military secrets. Let's just hope that some of the flying saucers being reported are really the latest US spaceships and not from some alien world."

Henry laughed, "Maybe for once, Europe will come and rescue us."

"Be real. Does anyone really think that NATO will be inclined, or be able, to come to our defense in a showdown?"

"Uh … I would hope so. But, you have asked a question more suitable for another program. Thank you, Dr. Peter Kinney. Thank you for coming all the way to New York. You have given us a lot to consider."

"You are welcome. Thank you for having me."

"This is Henry Haven-Harbinger signing off. Until next week at the same time—have a great weekend."

Villa

A worried frown crossed Tom's brow as he reached to switch off the program.

Just then, Kelly returned from their kitchen with two plates. She smiled and set them down on a table.

Tom hesitated. His cop's sensibility was now flashing red. "Did you hear that?" he asked carefully, hoping she had not heard.

"No, I was busy in the kitchen."

"Well, then, never mind," said Tom. "It was just science-fiction stuff. Ain't gonna happen," he said, picking up a sandwich and turning to her. "Okay, girl, I'm ready. Tell me what the heck was going on with that phone call to Mike?"

Chapter 3

Resort Island Roads

The roads on this island had not been repaired since they were first laid out. Progress was slow and bumpy for two late-model luxury vehicles, a van and two rental trucks making their way up an abandoned mountain road. Inside the lead car five passengers hung on tightly and called out to the woman at the wheel, begging her to slow down.

Colonel Rhee Su-jin merely gripped the wheel and pressed harder on the gas pedal. "Silence!" she ordered in their native language. She had no patience with weaklings and complainers. She had a vital job to do. Failure was not an option. Today they would conduct their first field test, the culmination of months of isolation and careful preparation.

Su-jin was the young commander of a field force entrusted with a top-secret mission for the People's Army. Not many were allowed outside the country, but she had proven her mettle many times.

As a student she had stood out from her classmates and was chosen to go to Switzerland for her degree. Next she entered the army officers' training corps, graduating with the rank of 2nd Lieutenant. She rose rapidly through the ranks having excelled in highly secret missions, until she was noticed by Dear Leader. Shrewdly, he recognized her as someone he could use, a smart, ambitious and capable young woman who could pass as a westerner. He quietly watched her career for some months until, satisfied, he called her into his headquarters for a personal interview. If she lived up to expectations, a complete examination and further training would follow.

Su-jin had no idea of the lofty plans in the works nor the part she would play, but she had worked hard to excel in every grueling test.

On the day she was called before the leader, she could not imagine what she had done wrong nor what her fate might be. Before entering his presence, she was forced to surrender her weapon and submit to a pat-down. Somewhat shaken and feeling naked, she stood at attention before his desk and saluted smartly.

He returned with a casual salute. "At ease," he commanded. "You have done well, Lieutenant Rhee, and are to be commended."

"Thank you, sir," she responded as she assumed the at-ease posture and directed her eyes to a spot below his desk.

"We believe you are ready for a very special mission. This may require you to be away from your friends, family and fellow officers for several months. It will be a difficult assignment, but if you do well, I will personally recommend you for a promotion. Are you willing?"

Su-jin was taken aback to be asked. She was always *commanded*, never asked. Her anxiety level soared. A bead of sweat broke out on her forehead. Nevertheless, she responded promptly, "Yes, sir." After all she could hardly say "No."

"Very well, Major. You will be in charge of a small force assigned to a post on an island off the northeast coast of Honduras, in Central America."

Su-jin stifled a gasp. He addressed her as Major. *Was that a slip?*

"You and your force will be undercover, posing as tourists. You will stay in a luxury resort, drive expensive vehicles and enjoy all the amenities. You will be given three months of preparation to assume these roles, as well as training in your real mission, which you will carry out in secret. When you leave here, you will go immediately into isolation for the purpose of training with your team of soldiers, so that no opportunity for a leak will be available. My aide will fill you in on more of the details, and escort you to your new assignment. Congratulations, Colonel Rhee, I am confident in your ability to carry out your mission. Continued success in your assignments."

He motioned to his aide. "Please show Colonel Rhee to her new quarters."

"Yes, sir," said the aide, rising from his seat and saluting the leader. "Come with me, Colonel Rhee," he said to Su-jin.

She saluted the leader, turned and followed the aide, somewhat shaken, but marveling at what just happened.

Training Camp

Their three months of training away from home, had been rigorous at times, relaxed at times. Her men all had scientific backgrounds. Studying the science and learning how to assemble and operate all kinds of ballistic missiles had been easy compared to the culture shock of learning to behave like wealthy western tourists. Colonel Rhee's group spent days shedding their military bearing, wearing jeans and tropical shirts or swimsuits and baseball caps. They ate hot dogs, chips and hamburgers, drank beer and learned how to order in Spanish and English.

It was hard to know when to frown and when to laugh at western jokes. They simply did not get the significance. They learned how to play baseball and toss a frisbee.

Snorkeling lessons had to take place in a large pool, since they were nowhere near an ocean. Su-jin hid her terror of the water fairly well, but she simply could not get the hang of the scuba lessons. And so it was decided that snorkeling was good enough. When they reached their destination, the men could scuba-dive on their own while she spent her days snorkeling or sun-bathing.

Their group had been selected, in part, because they had studied either Spanish or English as a second language. During their three months in training, they had to speak entirely in those two languages. Occasionally, someone would break out in frustration, swearing in their native tongue. Punishment was rapid and memorable.

Another selection criterion had been their appearances. In facial features, body-build and coloring, they must not give away their ancestry. Minor alterations to their facial features were performed the first week of their training, to be completely healed by the end. Su-jin received modest but perky breast implants. The men endured appropriate tattoos. The last week, Su-jin was given a beauty makeover, with lip plumping, eyelash implants, eyebrow-lift, a feather cut and blond highlights in her hair. The final touch was bellybutton and ear piercing and gold ear and belly rings. The transformation was complete. Su-jin looked amazing.

Meanwhile, fake passports and visas were obtained. Each member had to memorize his or her background and grow accustomed to the westernized version of his or her name. Colonel Rhee Su-jin had become Mrs. Sue Lynn Reese, recently widowed and childless office worker from San Francisco. Supposedly, she had taken leave from her job to return to college and would be down on the island enjoying a break.

The others were equally Americanized. Although staying in the same resort, they would not know each other until they met there and became acquainted. They would be careful, however, to avoid being seen together in a group, and to make friends with other vacationers, as well.

Practice Drill

Weeks later, having arrived separately on the Honduran island and having set up shop, it was only at prearranged times that they would make their way, singly or in twos, to the lab. Likewise on those rare occasions when they drove somewhere together they spread out to different pick-up sites, so they would not be noticed getting into the same vehicles. Once assembled behind darkened windows, they were relatively safe from scrutiny. Sometimes, these precautions seemed overly cautious to the point of being ridiculous, but Colonel Rhee was a stickler for discipline. She had been warned about the American CIA with their helicopters, spy-satellites and highly technical snooping devices.

Therefore, when they completed their field test today, they would break everything down, and remove every tiny trace of their presence from this site. They would use camouflage and try to be hidden, but if anything was noticed, the GPS coordinates might be available to local police who would swarm the site within a few minutes. Rhee hoped that they could make their escape before any enemy arrived.

Today was a practice drill. The unassembled medium range portable launch platform had been removed from the shipping container and placed in the small rental trucks. Their job, today, was to assemble it on-site, disassemble it and pack it back into the trucks before nightfall. Colonel Rhee would take her crew through this drill for as many days as it took to get this down pat. Only then would they attempt to place the missile on the launcher. For now the intermediate-range ballistic missile, IRBM, capable of reaching the US mainland, would remain packed in its wrappings, tucked away in the tightly locked shipping container.

The group started out in good humor, relieved to be doing something at last, after months of preparation. Within a few hours, tempers had already started to fray as parts were missing, misplaced or damaged from the transit. "Who the hell packed this mess?" screamed Su-jin in frustration, looking at first one man and then another. Everyone looked away. "Never mind, get to work and fix it." She quickly divided them into three groups. "You, you, and you," she pointed, "let me see who can make sense out of this the fastest." She held up a stopwatch, ready to click it. "Time begins. Start now!"

Her men were among the best. They huddled in their groups discussing options. Soon they joined together, nodded in agreement and swarmed over the arrayed disconnected parts. Su-jin stood back and watched her men scurrying this way and that. Within minutes the launch platform began to take shape before her eyes.

Three hours later she clicked the stop-watch and announced, "Three hours and six minutes. All right, men, let's see how fast you can take this apart and pack it away. Bear in mind, you need to do this correctly so that it will go together easily next time. Perhaps you need to follow the system we practiced back in the Homeland," she added, with a touch of sarcasm. Once more she held up her stopwatch. "Ready, set, time," she clicked the watch.

Two hours later the entourage was bouncing down the mountain road again. Before nightfall, they would move everything from the trucks back into the shipping container and return the trucks to the rental agency. Tomorrow they would take a day off to shop for better packing materials. Clearly, next time, the components needed to be wedged in tighter to withstand the pounding on the island roads.

Next Day—Beaching It

Her team had their orders. They were well-trained, smart people. Su-jin could relax and pretend she was a tourist. After all, it was part of her cover. She needed to be seen playing her part.

She found it impossible to sleep in. And so she was up in the morning with nothing to do. *I might as well enjoy the beach while I have the chance*, she thought. She had flown over the ocean but never been close to the actual water.

She was acting as Sue Lynn Reese now. She'd have to remember that in case someone said hello and asked her name. These Americans were so overly friendly. They would just come up to perfect strangers, look them in the eyes and start talking. She could not make herself do that. She could barely look someone in the eye, but with practice she had learned to return a smile, if forced. Life had taught her to never trust anyone, neither strangers nor friends, and so today would be hard for her.

Among her clothing she had been issued a bathing suit and some other items, packed into a bag marked Beach Wear. Su-jin pulled the bag out of the bottom of her trunk and laid the coordinated items on the bed. There was a roomy dress garment with large gaudy flowers in similar colors. The floppy brimmed hat and sunglasses, she understood. Even the strange footwear which matched the hat made some sense. But undergarments, made of the same material as the dress, were so skimpy they must have been on the bargain counter.

Su-jin put on the undergarments and looked at herself in a full-length mirror. She tugged here and there, trying to cover herself up. They just barely covered her private parts and certainly did not give any support. She pulled the dress garment overhead. This felt a bit more modest, but even this garment was so short, it left her thighs hanging out. She fiddled with the beach shoes for a few minutes trying to figure out how to put them on. Finally getting the hang of it, she looked down and wiggled her toes, deciding she rather liked the fake flowers on top of her feet. They matched the flowers on her dress.

She glanced at the glamourous "stranger" she saw in the mirror, picked up the huge beach towel in a matching color, wrapped it around her waist and left her room following the arrow signs marked Beach.

Several people preceded her toward the fence. Su-jin hung back, not wishing to appear friendly. And so the gate had swung closed as she approached. Su-jin reached for the handle giving it several pulls. Mystified she looked around for another entrance.

A man's voice said, "Here, let me help you." Surprised, Su-jin retrieved her hand and took a quick step back.

"Sorry, if I startled you," he said with a genial smile.

She watched in silence as he reached out a hairy, muscled arm, swiftly inserted his key card with one hand while reaching up to lift a locking lever with the other. He held the gate open for her and gestured for her to enter. "Please enjoy your day," he said.

Su-jin tried to smile and nod. Her lips choked on the words thank you, and so she pulled her towel tighter and walked ahead toward several rows of blue-painted flat benches. Some were set inside matching blue tent-like covers. Eyeing other guests, she chose one off by itself and sat down to observe. She was intrigued by the way people seemed to be clad in a variety of shocking garments no more ample than the underwear she was wearing. She had never seen so many naked bodies outside of the communal showers back home. Nearly all of them seemed overfed and under-exercised. She marveled at the variety of colors of skin, from bright red to dark brown. In pairs or family groups the women were either laid out on beach towels exposing themselves or rubbing oil on their bodies. Children ran up and down the beach, splashing in the waves or playing in the sand. A few athletic types were running on the shore, apparently training for something. Others were strolling.

A few young men and women were tossing a flat disk back and forth, seeming to enjoy chasing it as it slithered away in the breeze. Su-jin remembered practicing with a frisbee during her training. She moved aside as one flew her way and landed near her feet. *Could I throw the frisbee* she wondered briefly and then quickly rejected the idea.

A bronzed man sauntered her way and smiled through perfect white teeth.

Eyes downcast, she noticed his hairy legs, brief bathing trunks, and ripped abs.

"Excuse me," he said as he bent to pick up the frisbee. He waited for her answer. When she said nothing, he asked, "Would you care to join us?" He was the same man who had opened the gate for her.

Su-jin shook her head.

"Well, feel free to change your mind." He tossed the frisbee and wandered off.

Su-jin reached in her pocket and discovered an English novel with a shocking cover. Opening it she pretended to read while she watched the fascinating man from behind her dark glasses. He was taller than the men back home, with a beautiful lithesome body, relaxed and graceful as he moved.

Once again a frisbee landed near her feet. The man turned toward her. Quickly Su-jin hid the embarrassing novel. *Should I?* She rose from her seat and bent to pick up the frisbee, trying to hold on to her towel. As she moved to make a toss toward the man, the towel gaped open exposing her legs. Su-jin quickly grasped it closed. The bronzed man leaped into the air to make the catch and smiled, "Thank you, great toss."

Su-jin hastily retreated to her chair and settled down to watch. Meanwhile, one of the attendants approached. "Excuse me, miss," he said. "Sorry to bother you. I assume you are one of our guests."

"Uh," said Su-jin, uncertain of what to say. "Guests?"

"You're staying here?" he gestured toward the hotel.

Su-jin nodded.

He held a clipboard in one hand, poised to write. "Room number?"

"Room number," echoed Su-jin.

"May I have your room number?"

"Yes," she said.

He looked at her, waiting.

She fumbled in her pocket for her key card.

"No lady, the number is not on the key," he said and then he surmised she did not understand English. "Cuál es el número de tu habitación?" he asked.

"Oh, sorry. It's 513," she said in perfect English.

Puzzled, the man said, "I'll get your cushion." He trotted off to a pile of lounge cushions and returned with a large blue one under his arm. "Stand up a moment, please."

He quickly arranged the cushion, raised the back of her lounge chair and lifted the tent-shade of her cabana. "How's this?"

Wide-eyed, Su-jin nodded.

He hesitated just long enough to realize she was not going to offer him a tip. "Just let us know if you need anything," he said. "We have menus, all kinds of beach toys, surf-boards and snorkeling gear up in that little booth with the blue roof. We can serve food and drinks from the restaurant, arrange guided tours, boat rentals, fishing trips, luaus or show tickets at a discount." He pointed toward their booth and then moved off to the next customer.

Su-jin had curled up and closed her eyes for a few minutes, when the frisbee group broke up. Suddenly she felt a weight fall onto the far end of her seat cushion. Her eyes flew open. She gasped and sat erect bumping her head.

"Hi, there," said the bronzed man. "How ya' doin'? Enjoying this beautiful day?"

Su-jin gasped, drew her feet up to her chest and scooted back under her tent cabana as far away from him as she could.

"Oh, I'm sorry, did I frighten you?"

"N-no, not at all," said Su-jin, recovering her composure.

"Are you sure? Are you all right?"

"I'm fine. Just give me a second." She tried to relax.

"Sure, no problem." He grasped one end of the white towel around his neck and wiped his face, neck and arms. "I'm all sweaty," he explained.

38

Su-jin smiled, "Not at all, you're fine."

After thoroughly wiping his hands, he stuck out a paw, "Hello, I'm Steve."

Gingerly she took his hand and said, "Hello, Steve. Nice to meet you," just like she had practiced in training. And then she remembered to smile.

"Thank you, Miss … uh … Miss?"

"My name is Mrs. Sue Lynn Reese."

"Sue Lynn, I like that. It suits you, but as to the Mrs. part…I don't think so." He smiled holding up his left hand and wiggling his fourth finger. "I don't see a ring on your finger, Sue Lynn. In fact I don't even see a white spot where there has ever been a ring."

Nowhere in her basic training had they taught her about flirting. She didn't know how to take this man, or how to respond. "You are being impertinent, Mr. Steve."

"Not at all, Mrs. Reese. And it's Mr. Spalding, never married, but you can call me Steve." He leaned back, smiled, and made himself comfortable, making no move to leave. "Beautiful day, isn't it, Sue Lynn?" he asked.

She only nodded, which Steve did not see, as he was gazing out at some sailboats in the ocean.

He turned to look at her. "Have you ever been sailing?"

"No."

"Would you like to go?"

"I doubt I could do it."

"I'll take you," Steve offered.

"Oh my, no, I couldn't."

"Sure you could. It's easy."

She shook her head and looked down. This man had her flustered.

"Do you run?" he asked, fishing for an opening.

Su-jin was puzzled. *What did he mean by run?* This English language was so difficult. Words had so many different meanings. "Which run?" she blurted out.

Steve held up his palm and walked two fingers across it. "You know, run on the beach," he said, waving at the beach.

"Oh, run on the beach," said Su-jin, laughing at herself.

"You are pretty when you laugh," said Steve looking directly at her.

This comment made her blush. She understood this comment all too well. She looked down at her clasped hands and said nothing.

"Well," said Steve, trying again, "do you walk?"

"I guess so. Doesn't everyone?"

Reaching for her hand, Steve said, "Come along with me. We'll walk on the beach." He tugged on her hand rather firmly.

Su-jin jerked it back and used it to assist herself up off the chaise lounge, still grasping the towel. She reached for her floppy shoe with one foot. The shoe wouldn't go on without using her hands. She let go of the towel with one hand and it started to slip.

"You won't need those," said Steve. "Bare feet are best for beach-walking."

"Oh." She wiggled her feet in the sand.

"Shall we?" asked Steve, holding out one hand.

Su-jin took his hand.

"Why don't you leave the towel here to mark your place?" suggested Steve. "That way no one will steal your seat."

"Oh my, would someone do that?" She retrieved her hand, grasping the towel once more.

"Well, most people would assume that the seat was empty."

"I guess I was afraid someone would steal my towel."

"Really?" he asked doubtfully.

"Uh ..."

"It is a nice towel," Steve conceded.

"It matches."

"So it does."

Su-jin folded the towel in military precision and laid it squarely on the end of the chaise. Thinking better of it, she picked up her flip-flops and tucked them out-of-sight under the towel. Standing back she looked at them and frowned a bit.

Steve watched her, raised one brow and pursed his lips sidewise.

Turning toward Steve, she stepped forward toward the water, leading the way.

The waves were interesting to her. The way they flowed up onto the wet sand and retreated left a trail of foam which quickly disappeared into the sand. At first she moved aside to avoid getting her feet wet.

"Like this," said Steve swishing through the wave.

Su-jin copied him, giggling at the sensation.

No one owned the beach. People from neighboring resorts mingled. Neither Su-jin nor Steve had any idea that fate would soon bring them together with the young honeymooning couple from Carson City, New Mexico, running past them now.

Chapter 4

Kelly fell behind in their race to the blue roof. "Hold up," she called out to Tom. Not hearing, he raced on ahead as if he could go on forever, barely exerting himself.

Out for a morning run on the beach, the honeymooners had decided to race to the hotel with the blue roof. Neither one took any particular notice of Sue and Steve, who were just one of the many couples walking on the beach.

"Tom, wait!" Kelly came to a halt, hands on hips, chest heaving.

Tom turned around and pedaled backward scanning a cluster of beachgoers blocking his view of her. He reversed course and trotted back to find Kelly trying to catch her breath. "Hi, babe, you're looking a little winded."

"Yeah," she said breathing rapidly through her mouth.

"Want to take a break?"

"Yeah," she nodded.

Tom eyed her with a bit of concern. "Okay, let's grab a seat over here. I'll pick up a couple of drinks, okay?"

"Thanks," said Kelly.

Tom led her over to a seat and helped her down. "I'll be right back."

He walked over to the beach shack with the blue roof. All the cabanas were blue, as well. Each resort hotel used a different color, to make it easy for their guests to find their way home after they wandered away up and down the open beaches. "Hi, my wife's a little winded," said Tom. "We're from Barefoot Sands Resort, down the beach." He gestured toward the direction. "May we use your beach chairs for just a couple minutes until she recovers?"

"Certainly, sir. Help yourself."

"May I buy a couple of waters?"

"Sure can," said the man, "Lemon, peach, raspberry or strawberry?"

"Anything, just so it's cold and wet," said Tom.

"Here you go," said the attendant, as he popped the tops on two ice cold drinks. "It's on the house for honeymooners," he said with a knowing smile.

"Are you serious?" asked Tom. "I have some US money right here." He reached for his wallet.

"No, sir, we don't take cash. We're only set up to put it on the hotel bill for our guests."

"Well, thank you so much. That is very kind." He hesitated a minute, "You said honeymooners. Was it that obvious?"

"Just a guess. We get a lot of them," he chuckled. "Might be a good idea to give the new wife a bit more rest, you know?" he suggested with a twinkle in his eye. He turned toward his work and started humming.

"Well, thanks for the drinks and the advice," said Tom.

"You're welcome, sir. Enjoy your day. It won't get any better than this."

Returning to Kelly, Tom sat down beside her and offered her a bottled water. "Have some of this, honey," he said.

"Thank you." She took the bottle and drank. "That is so good."

Tom drank from his bottle. "Hits the spot."

She glanced at the blue shack. "Do they mind if we rest here for a few minutes?"

"Not at all. In fact they gave us the water. Wasn't that nice?"

"Really?"

"Man suggested I might be tiring you out."

"Hmm." She shrugged.

"Will you be okay to walk back? We can take the shuttle if you want."

"I'll be fine. Just give me a couple minutes."

After ten minutes, they began their walk back to their villa. Tom made a point to stop and rest twice.

Rental Scooter-Bikes

Kelly had had a restless night, dreaming about bad men trying to catch her. This had to stop. She and Tom made a plan of sorts.

After lunch at the pool, they grabbed a ride on the shuttle that ran up and down the beach road every half hour. Getting off at the Blue Roof Hotel, they strolled to a nearby scooter-rental shop.

"Two bikes," said Tom holding up two fingers.

"Cuatrocientos lempiras," said the clerk.

"Do you take US dollars?"

"Twenty dollars an hour."

"All right. We'll take two bikes for an hour, maybe two."

"You pay two hours, now. Bring back early. Get refund," he smiled showing a space between his teeth. His wrinkled black skin folded into a dozen creases.

"I'll pay you for one hour and then we'll see."

"One-hour hokay. Deposit twenty dolla. Bring back. Get refund deposit."

This guy drove a hard bargain. Tom smiled and reached for his wallet. "Here you go. Twenty for the deposit. Twenty for one hour."

"ID, por favor."

Tom fished his driver's license out of his wallet. The clerk stepped into a small open-air office and ran the license through a small machine. Turning his back to check the number against a crinkled list of numbers, he shook his head ever so slightly, punched the transaction into his cash register, opened the cash drawer, deposited Tom's two bills, closed the drawer and tapped a button for the printed receipt.

Returning to Tom, he handed Tom the keys and receipt. "Gracias. Drive carefully. Enjoy."

Tom helped Kelly onto the seat of her motor-scooter. A quick demonstration covered the simple controls. "Are you okay, babe?"

"Got it," said Kelly, pulling her cap on tight and adjusting her dark glasses. "Put it in neutral to start." She wiggled the gear shift lever with her left hand. "Stomp on the starter with your left foot." She revved the motor. "Shift into forward. Give it some gas and push off." She took off down the driveway and swung out into traffic with Tom close behind. "Whee," called Kelly grinning and waving at Tom with one hand while she steered with the other.

"Don't forget to brake with your right foot," called Tom into the wind, with a worried look.

Forty-five minutes later they had ridden to the end of the island and back. Kelly pulled into a small café with outdoor tables covered by cheerful umbrellas. "Thirsty?" she asked.

"Sounds good," said Tom.

"How's this?" she asked keeping her head down and choosing a table next to a low stone wall topped with flowing vines and mixed tropical flowers.

"Perfect," said Tom, pulling out a chair for Kelly and then seating himself. Feeling a bit uneasy, he eyed the security camera and adjusted the umbrella to block its view. "Hungry?" he asked, picking up a menu.

"Not really. Just something salty and a tall drink."

Tom signaled a barefooted waiter dressed in white shorts and an untucked tropical shirt.

"Saludos," said the waiter. "Su orden por favor" (Your order, please.)

"Ponche de frutas y patatas fritas," (Fruit punch and chips), said Tom holding up two fingers.

The waiter nodded in understanding and hurried off.

Soon two tall frosty drinks appeared with a bowl of tortilla chips and dip. Tom offered some local bills.

"Try these," he said to Kelly. They would be careful to pay with local currency and speak Spanish, reserving English for each other in quiet tones when no one was near.

"Mm, delicious," she said. She chewed on a chip and sucked her drink through a straw. "If I'm correct, The Pigsty gift shop is on the next street over."

"Ya' think?"

"Well, I remember that hotel across the way where I got the taxicab." She pointed. "And I ran through a kind of store and out the back. It could have been that one there, but I'm not sure."

"Are you sure you ran out the back? That looks more like a front."

"You know, you're right. Maybe it was the front. I was so scared."

"Well, you stay here while I go and investigate. You'll be safe here, I think. Keep your phone handy in case you need me." Tom took out his phone and turned it on. He slipped it into his shirt pocket and patted the knife-holder, attached to his belt. He rubbed the back of his neck with one hand, trying to wipe away the creepy feeling of being watched. Taking one more glance around, he bent to kiss her cheek. "Be really careful, okay?" he whispered in her ear.

"Always, sweetheart, you know I will. You be careful, too." Tom nodded and hurried away.

Kelly tucked her small cap and glasses away and pulled a huge pair of sunglasses out of her new backpack, followed by a floppy cloth hat that she pulled down around her face. A well-thumbed paperback novel in Spanish completed her makeshift disguise. Although she was out in the open she sat with her back toward the street, in the shade of the umbrella, hiding in plain sight. Her body was covered by the table and the solid, high-back wing chair blocked the view from both sides.

Off in one corner of the patio, well-hidden behind a tropical plant, a tall bronzed man watched these proceedings with interest.

Kelly's only personal weapon was a sturdy pen-like device made of a strong space-age material which passed through airport security. The sharp point could be lethal in close quarters. Opened, it concealed several handy tools useful in cutting ropes and picking handcuffs. It even hid a small breathing tube for use underwater. Her older brother, Mike McBride, had taught Kelly how to use it when he first became a cop while she was still a kid. She had practiced with it until she became proficient. Still living nearby, Mike also encouraged her to take martial arts training along with him. She was never as good as he was, but she knew enough to protect herself.

Tom had bought inexpensive metal knives for both of them after they arrived here. They would not be able to take them home, of course. Kelly checked the bottom of her pack to make sure the knife was still there. She clipped her pen weapon onto the cover of her novel and turned a page.

Tom mounted his motorbike and sped away. He would approach the side-street from a different direction. Meanwhile, Kelly did a convincing job of reading her novel while constantly scanning the area from behind her dark glasses. She sipped on the drink and nibbled slowly on the chips. When the waiter approached she shook her head and turned back to her novel. She presumed the bad guys had alerted the waiter to watch for an American woman of her size, but that could describe a lot of the female tourists. Unless they offered a large reward, they would not garner many leads. Whereas, if a reasonable sum was offered, they might have more false leads than they could handle. Kelly felt safer with each passing hour. In time the search would stop. Surely, her discovery and subsequent escape would become old news, wouldn't it?

The tall bronze man hidden behind the tropical plant had already snapped several shots even before Tom left. Now he transmitted them to headquarters. He watched Kelly pretending to read her novel for fifteen minutes before he realized nothing was going to happen. There was only one of him. Too bad he couldn't have watched them both as Tom's trip was more interesting.

Tom approached The Pigsty from the other direction as a precaution, not that it probably mattered. He drove by on his scooter and circled the block to drive by again. This time he turned into the alley behind the buildings and approached The Pigsty from the rear. He saw a large shipping container and two vehicles parked next to the back entrance. Now would not be a good time to snoop inside that container. He would come back when there were no cars around.

Back at the restaurant Kelly waited until she heard his signal. Then she quickly packed her things into the backpack, mounted her motor-scooter and took off to follow Tom. She couldn't wait to hear what he had learned.

Checking Clues

Back at the villa they changed into their beachwear and found a secluded spot where they could talk. They spread a blanket on the sand and plopped down. "Okay, spill," said Kelly, eager to hear.

"Not a lot to tell, babe," said Tom. "I rode by the front of the store a couple times and snapped a few pictures. It's exactly as you remember, the tall decorative fake front, the pig display in the window." As he talked he scrolled through the pictures on his phone. "Here's the front," he said and handed it to Kelly. "Now scroll through the next few shots."

Kelly nodded. "Yes that's the front. What's this next shot?"

"That's the back."

"A shipping container?"

"Yup."

"Do you think it was a new one?"

"Keep scrolling."

"Looks like a close-up."

"Yeah."

"You got the lettering on the side."

"Can you make out any of it?"

"Strange looking characters."

"That's what I thought."

"Chinese, maybe?"

"Keep going," said Tom.

Kelly scrolled through the next few shots. "You got their cars and license plates. How did you do this? You promised me you wouldn't stop," she accused.

"I keep my promises, babe. All I promised was I would not get off the bike."

"So how ...?"

"I'm a cop, remember?"

"I still don't see ..."

"Do you know how many pictures of moving license plates I've taken in my career?"

"No."

"Would you believe hundreds, maybe thousands? A license plate can tell you a lot. This camera has a super-fast lens."

Kelly gave him a skeptical look and continued scrolling. "Probably those are rental cars anyway."

"Can still tell you a lot."

"Okay, detective, how do you plan to run those plates? I doubt you're going down to the local police station."

Tom just smiled. "We smart cops have our ways."

"Mm, yeah right." She gave him a playful swat. "You have your ways, all right."

"Stick around and I'll show you," he leaned over, stole a quick kiss, grinned and reached for the phone. "May I have that?"

"Here," said Kelly, blushing slightly as she handed back the phone.

Tom took the phone and began tapping some numbers.

"What are you doing?" asked Kelly, always curious.

Tom continued tapping. Then he stopped and watched his display. He looked up and grinned. "The Island Car Rental Agency, 5820 Shoreline Drive." He tapped a few more times and waited. "It's a sixty-day rental with option to renew for another thirty to ninety days," he paused and scrolled down, "to a Mrs. Sue Lynn Reese, 513 Blue Roof Hotel." He looked up with a satisfied smirk on his face.

"How'd you do that?" Kelly was incredulous.

"That advanced international computer skills class finally came in handy," he said.

"You hacked into their system!"

"Let's just say I borrowed some information."

"You're sly, Thomas Turbulo."

"Well, the cops down here don't have a lot of money to spend on software security."

"You're being too modest."

"Let's see what else we can find about Mrs. Sue Lynn Reese. Hm. She lists her home address as Encino, California. Not sure I can get in there. California has a much better system. If I was back home in the precinct I would have no trouble. Let me try, first, and then if I can't get in, I'll text one of my buddies."

Tom spoke into his phone and tapped away several times, pausing and shaking his head. Finally, he looked up. "We'll have to wait. I sent a query off to Lt. Sam Mulholland. He'll check it out when he has some time."

"How about Mike? Can he help us?"

"That's a good idea. I hadn't thought of Mike. He can run her name through the known terrorists lists and maybe come up with something."

"Why don't you send him those pictures, too. You know, especially of the words on the shipping container. He can probably decipher those for us."

"Great idea. I'll do that, too." Tom worked his phone for a couple minutes and then he stood up. "What say we walk back to the villa, babe, while our people back home go to work? I'm hungry." He held out his hand to help her up.

Kelly took his hand and started walking. "I'm going to be ready for some pool time and then maybe a nap."

"Yeah, let's do it, only nap first and then maybe pool," he wiggled his eyebrows and grinned.

In the distance a tall bronze man lowered his high-powered binoculars, put them into their case which he stowed away in his pack. He walked to his motorcycle parked behind a sand dune, mounted up and sped away.

Carson City PD

Lt. Samuel Mulholland was surprised to receive a message from Sgt. Turbulo. "Good grief! Isn't the man on his honeymoon? Well, let's see what the heck he wants," muttered Sam under his breath. It didn't take long to query the California records for a Mrs. Sue Lynn Reese, approximate age twenty-five, at the Encino address. There was no such address or person and never had been. For good measure he checked a few nearby states and the federal database. Nothing popped up. Sam typed a reply to Tom with a little note adding, "Take my advice buddy and pay attention to your current wife. Ignore all other women."

D.C. Senate Office

Senator Mike McBride was at his desk when the message came in from Tom Turbulo. He read the letter and studied the pictures. Unsure where to start, he signaled Cynthia Patterson-Eastman, his personal assistant, bodyguard and chief of his investigating squad. "Please come into my office, Ms. Patterson."

"Yes, sir," she answered.

Mike looked up as she entered. "Good day, Ms. Patterson." Cynthia was married to Major Sky Eastman, now, but kept her maiden name professionally.

"Good day, sir. How may I assist you?"

"Take a look at these pictures, please, and tell me what you see."

Cynthia studied the shot with a detective's eye. "This first picture is taken from a street, looking toward a storefront of some kind. La Pocilga means pigsty in Spanish. Odd name for a store. The address is in Honduras. Hard to tell what's in the window. Too much reflection. Probably taken in the late morning." Cynthia studied a moment more and then looked at the second picture.

"This is a long shot of either a different building, or the same building from a different side, maybe the back as I see a dumpster, also a couple of vehicles and a shipping container."

"According to Tom's message it is the back of the same building," said Mike.

"Tom?" queried Cynthia, looking up.

"Sgt. Tom Turbulo," answered Mike.

"Your new brother-in-law?"

"Right. Just married and already in trouble."

"Honestly! What the heck are they up to now?"

"Well, yesterday Kelly was attacked while snooping around inside that gift shop. Whatever she saw was connected with the gift shop, I guess."

"Outside?"

"No, she must have found another whole operation going on hidden inside or behind the store somehow. She thinks it is some kind of terrorist group or something. I don't know."

"So Tom went back today to take these pictures," Cynthia surmised.

"Right."

"Yeah, these license plates show it is in Honduras, and the make and model of the vehicle is typical of rental agencies in that area. I'm guessing it's one of the tourist resorts."

"Yeah, I'm pretty sure you're right."

"Interesting writing on the shipping container. May I enlarge that?"

"Sure, go ahead," answered Mike.

Cynthia used her fingertips to zero in closer on the writing. She furrowed her brow, "This is in Korean."

Mike gave a low whistle through his teeth. "North or South?"

"Can't tell until I look it up. Korean is pretty much the same, north or south. This string of routing numbers will tell us the origin, destination and the ports of departure and entry. This other sticker will be the identity of the ship that transported the container."

"Can you look that up?"

"Yes."

"What else do you see?"

"Well, judging from the shadows, these shots were taken about the same time of day and in the opposite direction from the first picture."

"That fits," said Mike.

"The cars seem to be dusty."

"They've been out on some dirty roads."

"Yeah they have and recently I would judge." She squinted and looked closer. "I can check out those license plates, if you want."

"No, Tom has already done that."

"Sneaky."

"Says both cars were rented four weeks ago, for sixty days using a fake identification."

"Are you kidding me? He found that out already?"

"Yeah, a Sue Lynn Reese, age twenty-five, of Encino, California."

"Well, I can double-check that."

"Good idea. But, I'd rather you traced the shipping container."

Cynthia examined the picture again. "The shipping container is fairly new."

"Oh yeah?"

"No visible rust. Also, it hasn't sat here very long."

"Really?"

"Look at this picture." She handed the iPad back to Mike. "See the weeds?"

"Oh yeah, they aren't very tall, are they? Good eye, Miss Patterson."

"Thank you. I'll need to do more research on those numbers and words."

"Thanks, please do that as soon as you have time."

"I'm on it, boss," said Cynthia, rising from her chair.

"Before you go, I already talked with Jo Rench over at CIA."

Cynthia eased back down in her chair. "I hope you didn't give him all this stuff?"

"No, I talked to him yesterday, after I got the panic call from Kelly."

"What call was that?"

"She called me, almost in hysterics, saying she had discovered something awful. Then she hung up. I figured out her location and called Jo Rench to see what was going on down there."

"And he told you?"

"Well, no he didn't say."

"Well, Senator, so far as whether you should report this to the CIA, it depends on how much you want them to know about your sister's activities."

"I just want to ask them what's going on?"

"They will take your information and they won't tell you a thing."

"I see. I guess you're right. But I feel like I need to do something."

"Let me check for you. I have friends in the CIA and other places." She rose again.

"Thanks, Ms. Patterson," he said with some relief.

"One more thing," she added. "Is it all right if I send a couple of my staff down there?"

"Okay."

"Expenses paid?"

Mike laughed. "You got me there. Okay go ahead. Send them off to the beach, on the office expense account." He sighed and turned back to the papers he was studying.

Chapter 5

Reports and Options

On the other side of the world, Dear Leader, the chairman of Homeland, had been wearing a smiling public face and signing non-proliferation agreements with various world powers. At the same time, his atomic scientists, physicists and engineers were working twelve-hour shifts, seven days a week on a vital and highly secret project, well hidden from even the latest state-of-the-art satellite cameras.

Whispered rumors in back rooms had long hinted at the possibility that the Chinese, Russians and the US might be secretly working on tactical light-weight atomic weapons in violation of their treaty, but politicians doubted if such a thing could happen or would even work. It was impossible to separate fact from fantasy.

Meanwhile, Dear Leader's NR&D (Nuclear Research and Development) complex was buried deep within a mountain. Experiments were conducted well out of sight and sound and had gone undetected for many months. If they succeeded, their country could make a nuclear weapon capable of destroying an entire advanced civilization with one tiny bomb, no larger than a few pounds. The prospects for world domination and enormous riches were intoxicating.

Isolation from the rest of the world made it possible to carry on in secret. Very few outsiders were allowed and those visitors were carefully controlled and restricted in their access to certain areas of the country. United Nations inspectors were only permitted into certain sites that Dear Leader disclosed. Reporters had been invited to witness a staged blowup of their "principal test site" when, in reality, it was one they no longer needed.

The rest of the world totally underestimated their prowess, thinking of the country as bankrupt, backward and uncivilized. It was simply not considered a major threat. This was exactly what Dear Leader wanted them to believe. Even the Chinese were fooled.

President Bigelow of the United States bragged about how his army was the strongest on earth and could easily flatten any opponent. That might be true in a conventional 20th Century war. But the vaunted US military would be rendered completely useless, unable to lift a finger. Dear Leader smiled to himself and contemplated how sweet it would be when the entire United States and most of Canada was in his sights. Of course the weapon had to have a delivery vehicle. That is where Su-jin and her mission came in, as did a duplicate force exactly like hers operating many miles away. Dear Leader always had a backup.

It was early morning when DL (Dear Leader) entered his office. After morning tea, he called General Lee in for their morning briefing. General Lee was in charge of 작전 핵겨울 (Jagjeon Haeg Gyeoul) or JHG for short, meaning Operation Nuclear Winter. Only two people knew the complete details of the operation, Lee and himself.

"Good morning, General," said DL.

"A fine morning to you, as well," said Lee, who had been kept waiting in the outer office while DL had his tea. Lee stood holding a large stack of folders under his arm.

"Please be seated," said DL. "and tell me how things are going with plans for JHG."

"Here is a one-page printout summarizing the activities," said Lee, placing the page exactly in the center of DL's desk. "I have details here," he said indicating the heavy folders he was juggling.

"You may set those down," said DL, impatiently waving at the corner of his desk.

None of DL's associates dared speak out of turn or take any initiative in his presence, lest they incur even the slightest disfavor. Demotion, imprisonment, torture and even death could result, depending on DL's mercurial mood at the moment.

DL glanced at the printout. "Never mind this, just tell me. How is the Honduras mission going?"

"Good news, Dear Leader. Colonel Rhee reports that the first practice attempt at assembling the portable rocket launcher was a success. It took five hours. They will continue practicing and hope to get that time down to about three hours."

"Are they being careful?"

"Yes, of course, by all means."

"I expect nothing less," observed DL stifling a yawn. "Have there been any leaks?"

"Our computer room reports that hacking attempts by hostile powers have been successfully blocked," said Lee.

"How about from friendly powers?" DL laughed at his own joke.

"None, sir."

"Very good, Lee. We need to stay ahead of their lazy security teams."

"Absolutely, sir. No one in the entire world can defeat our cyber-warfare offense and our defenses."

"Not even close?"

"No, sir."

"Good. Let's keep it that way."

"Of course, sir."

DL adjusted his short rotund frame in his oversized chair, "Now that we have sucked all the information possible out of their defense department computers, our people will forge ahead." He licked his lips and squinted at General Lee through his round silver-rimmed glasses. "How soon will the IRBMs be ready to launch?"

"Well, as you know, the missiles have been delivered to Central America and are hidden inside the shipping containers. Currently, both teams have begun practice-assembling the portable launch platforms. They will not attempt a practice launch until you give the order."

"I'm not sure we can risk a practice launch because of the likelihood of detection. It would give away their location and put our teams in danger," DL pointed out.

"You are quite right, of course."

"Well, then?" DL asked as he scanned the document in front of him.

"I see only two options."

"What are they?" DL looked up expectantly, almost daring Lee to make a suggestion.

"We could move the teams after the test launch, or we could try a test launch from a completely different location."

"Or we could simply omit any test launch, entirely," DL rejected Lee's ideas.

"Uh ...how would that work?" General Lee was wisely hesitant to question Dear Leader.

"Haven't we already done enough testing from here? You reported that we could easily reach Japan or as far away as Guam."

"Yes, sir, we believe that is possible, but ..."

"Well then, it's a matter of measuring the risk both ways," DL explained with uncharacteristic patience. "If we conduct a test launch from Central America we give away our capabilities, render the team useless and maybe even lose some of our team members. The stupid Americans think they are safe because we do not have a long-range ballistic missile capable of carrying a nuclear warhead. Our intermediate range missile tests lulled them into complacency, especially when we allowed a few of them to crash into the ocean." He chuckled and continued on.

"On the other hand if we omit any tests and simply go for the final launch, we take them completely by surprise. There is only a small risk of failure entirely." DL dismissed that idea with a shrug.

"I see what you're saying." Lee knew full well the penalty for failure. He had been ordered to administer penalties more than once during his rise to the top.

"Option two appeals to me for this reason, we have two separate teams on identical missions. If one fails, we still have the other." DL had made up his mind.

"That's true," Lee felt the pressure tightening. Cautious by nature, if he had his way there would be dozens of launch tests. This payload was a highly technical experiment and still under development. The idea of going forward without multiple tests was appalling to him.

"We only need one of them to succeed," DL barreled on.

"Chances would be even better if we had three identical teams," bargained Lee, "or a half dozen. Perhaps, the more we have the better."

"How many teams can you mount in thirty days?" challenged DL.

"Well, I don't know. Each team has taken years to select, train, equip and deploy," Lee admitted.

"And so there is a time delay element," surmised DL. "Also, more people to supervise and control," he scoffed. "More chances for leaks."

Lee remained silent.

DL paused for a full minute, reconsidering his options ... "Colonel Rhee will not fail," he decided. "Now, let's talk about the weapon. How close are we?"

"There are many elements to consider."

"For instance?"

"Well, the most difficult are the weight limitations."

"Explain, if you please."

"We know the payload capacity of our Intermediate Range Ballistic Missiles. And so that is the first limitation. The weight governs the distance."

"How so?"

"The lighter the payload, the farther it can go. By that I mean the rocket has so much power. The heavier payload requires more power to push it out. You figure the distance you need to go and scale your payload accordingly."

DL cut to the chase, "In other words I can throw a lighter ball farther than a heavy one."

"Exactly."

"Let's say you want to place a satellite over Kansas City at two-hundred-fifty miles above the earth at a certain speed, you would measure the distance, weigh your payload and calculate how much power that would take."

"Like I said height plus weight times velocity equals power needed."

"It's quite simple, if you haven't got enough power, then you lower the weight of your payload. Thank you, General Lee. That will be all."

Lee picked up his armload of files. "Good day, sir."

DL waved him away.

Lee backed out of the room and closed the door.

CIA Headquarters, Washington DC

Assistant Director Joseph "Jo" Rench sat at his desk going over the daily reports from his men in the field. Jo had charge of their Central American assets including the satellites, the men, women and their computers. Fortunately things were quiet right now, and so he had time to catch up on requests for favors from members of Congress. Yesterday he assigned one of his agents to check on Senator Mike McBride's sister. Jo pulled up the man's report on his iPad.

Officer #Q00059667JB

USCIA

Surveillance Report

Subject: Mrs. Kelly Turbulo, American citizen, age 23, Carson City, NM, vacationing Honduras.

09:13 AM Subject leaves Barefoot Sands Resort accompanied by husband Sgt. Tom Turbulo of CCPD.

09:15 AM Subjects board beach shuttle

09:26 AM Subjects disembark Blue Roof Hotel stop.

09:30 AM Subjects rent two motor-scooters.

Quiz bike attendant. N/A. Detect/photograph crumpled list of license numbers. None matching subject.

10:00-10:46 AM Subjects at local café. Husband leaves at 10:15 AM. Mrs. Kelly Turbulo leaves at 10:46 on motor-scooter. Returns to villa.

11:00 AM Observe both subjects together on secluded area of beach, reclining on blanket.

Awaiting further assignment.

***

Assistant Director Rench filed the report and dictated a quick answer:

To Officer #Q00059667JB. Continue surveillance of the Turbulo woman and the other subject. Report any suspicious activity.

Cynthia's Team. US Senate Office Building

"Good morning, everyone," Cynthia said, pulling out a chair at the head of the conference room table.

"Good morning," chorused six lovely ladies, seasoned operators on Senator Mike McBride's secret team of investigators, each one as highly skilled as a Navy Seal or Army Green Beret. In fact three of them were former military officers, as well as holding Masters' degrees in technical areas and languages.

"I have a juicy assignment for two of you lovely ladies," said Cynthia. "You will need to pack your beach-resort wear and prepare to be out of the country for no more than two weeks."

She was greeted with a chorus of "Oo's."

"Count me in," said Sharon.

"Me, too," Agatha said.

"Not I," sighed Cherry. "Kids have a soccer match."

Wilma and Dolores shook their heads sadly.

Marion said, "I can go if you need me, Cynthia, but I'll have to rearrange some things."

"All right, then," said Cynthia. "We'll send Sharon and Aggie. Marion can stand by as back-up. Cherry can go back to the front desk. Dolores and Wilma, you can finish your work this morning and then go on home." Dolores and Wilma were the two part-timers.

The three women had regular daily duties in Mike's office and so they left the room.

"All right, here's the skinny," Cynthia addressed the three remaining. "Senator McBride's sister, Kelly Turbulo, is honeymooning on a resort island off the coast of Honduras. She has stumbled onto some sort of mysterious operation which may have put her in danger. Your job will be to act as bodyguards, staying as close as possible without interfering in her honeymoon. Hopefully all will go smoothly without further incident and you will spend your time in fancy restaurants and relaxing on white sand beaches, until you escort the happy couple home to the good old US of A."

"All expenses paid?" asked Agatha.

"You betcha',"

"Sweet," said Sharon.

"I have your tickets and reservations in these packets, as well as pictures of the subjects and their contact information. I'll be working on this end investigating the suspicious group that attacked Kelly and caused Senator McBride's concern. Please report to me twice daily and more often, as needed. Any questions?"

"Can you tell us anything more about the attack?"

Cynthia filled them in on everything she knew. "Now, go get packed for the Barefoot Sands Resort, ladies. You'll be in the main lodge." She smiled and rose to leave the room. "I'll head back to my boring desk job."

Blue Roof Café

Su-jin moved through the cafeteria line pushing her breakfast tray. She spotted two of her men already eating at a table off in a corner. After selecting her food she moved to a nearby table, nodding briefly at her men while making no further recognition.

There was no need to pay for her food, as this was an all-inclusive resort. Her room key and number were sufficient for the clerk who was checking people in. This entitled her to three meals a day plus drinks and snacks.

Steve Spalding had been lurking in the lobby hoping to spot her again. Quickly he picked up a cup of coffee and approached her table. "Good morning, Sue Lynn," he greeted her with a handsome smile. Pulling out a chair he sat down at her table without asking permission.

Su-jin was a bit startled, at a loss for the correct words to tell him to get lost. Instead she nodded and began eating her food.

Steve commenced to chat about the fine day and the beautiful weather, inviting her to respond.

Su-jin merely mumbled agreement and continued eating.

From time to time, Steve cast his eyes around the room to see if anyone was watching. Two men nearby were gawking at them, wide-eyed. Steve allowed his eyes to pass over them without appearing to take notice. Clearly the two men had some interest in Sue Lynn. Steve leaned closer to her, to see what they would do. One of the men half-rose from his chair. The second man laid a hand on his arm to stop him.

"Can I interest you in another walk on the beach, after breakfast, Sue Lynn?" Steve inquired. "Or maybe you would enjoy joining us for a go at Frisbee."

Su-jin found her tongue. "No thank you Mr....uh...Mr. er, I'm sorry I forgot your name."

"Steve Spalding, at your service, Sue Lynn," he saluted, grinning broadly, keeping one eye on the two men.

Su-jin started to sweat. Her men would report her if they thought she was being traitorous. "Well, um, I'm sorry, Steve, but I have other plans today."

"What a shame, Sue Lynn. Are you free later, perhaps for dinner? I know a great restaurant."

"No, thank you. I'm really busy."

Steve took her hand. "Please," he begged. "Pretty please, with sugar on it." He made a point of petting her hand and saw the two men appearing more agitated.

Su-jin snatched her hand back and could not resist a guilty glance toward the two men. This was the "tell" that Steve needed. He was sure, now, that these three were associated.

"Well, dear, I'm surprised that you are so busy. Didn't you tell me you were on vacation, taking leave from your job at … uh … I can't remember. Where did you say you worked?"

"I said I'm an office worker on leave from my job, in order to return to college." Su-jin recited from memory.

"Oh yeah, now I remember," said Steve. "From San Fernando, right?"

Su-jin nodded.

Now, Steve could tell she was lying, because she had said San Francisco yesterday. Clearly she didn't know California geography since San Fernando is the name of a town, not a city. It is in the San Fernando Valley, closer to Los Angeles than San Francisco, and nowhere near the ocean.

Steve tried one more test. "San Fernando is such a beautiful city, right on the bay. Do you work downtown or up in the hills?" He took care to pronounce it clearly.

"Sometimes," Su-jin shrugged, unable to supply an answer.

"Don't you just love those little cable cars?" he asked.

"Right," said Su-jin, unsure of what they were.

Steve stood. "Thank you for your pleasant company. Perhaps we'll meet again tomorrow." He bowed slightly and pushed in the chair. "Good day, Sue Lynn."

Barefoot Sands Resort, the Villa

Kelly and Tom considered their next step. It was time to make a plan.

"Somehow, don't you think we need to get a look inside that shipping container?" asked Kelly.

"I suppose we could go over there later," Tom agreed. "But I think it might be best to go after dark."

"Wouldn't that be too obvious? I mean sneaking around after dark, with a flashlight, looks a bit too suspicious, doesn't it?"

"Point taken, honey, but if we go during the day, what if someone is there?"

"Well, they don't know you. Maybe you could go into the store and distract them while I check out the container."

"Them?" asked Tom. "How many of these people are there?"

"Not sure. Maybe a half dozen. I was too busy running to count."

"I'm not crazy about those odds," Tom remarked.

"Wait a second, I think that's my phone," said Kelly. She moved into the bedroom to fetch her phone. Returning to the couch, she frowned at the number. "I don't recognize that number. Who could it be?" she gazed at Tom.

"Well, honey, you won't know unless you answer it," Tom stated the obvious.

Kelly tapped the phone. "Who's calling?"

"Kelly, don't hang up," a female voice pleaded. "Cynthia sent us. We'll be right over."

"W-what?" Kelly stammered. "She says Cynthia sent her," Kelly informed Tom, who was equally mystified.

The voice grew louder. "You know the Cynthia that works for your big brother."

"Oh my goodness!"

"Yeah right. Now don't say any more. We'll be right over. Wait there. Give us ten minutes. Don't leave."

Kelly plopped down in a chair, her mouth hanging open.

"Who was it?" asked Tom.

"She said they'd be right over."

"Who will be right over?"

"Uh, I'm not sure. Some woman."

Tom rose and went into the bedroom to strap on his weapon. "I'll watch," he said. "Why don't you stay out of sight?"

A tense fifteen minutes later the sound of a moped approached along the boardwalk. Two lovely women were riding tandem. They pulled up to the front of the cottage, parked and approached the front door. Before they could knock, Tom stepped outside and closed the door, crossing his arms and putting on his tough cop grimace.

"Heh, Tom. How ya' doin'? I'm Sharon and this is Agatha." She held out her hand.

Agatha held out her hand, as well, "People call me Aggie."

Tom kept his arms crossed. "Nice to meet you Sharon. But, we're a little confused here. Who the heck *are* you, and how did you find us?"

Both women laughed. Agatha spoke, "Your brother-in-law sent us."

"What brother-in-law?"

"Senator McBride, doofus. Kelly's brother. How many brothers-in-law do you have anyway?"

At that point, Kelly came flying out the door and wrapped Agatha up in a bear hug. "For heaven's sake Aggie. How on earth?"

"We came by airplane, of course. How else?"

Sharon joined in a group hug. "Surprised?"

"Speechless."

"Well, we're here, now." Turning toward Tom. "And who is this handsome guy, guarding your front door, huh?"

"Oh excuse me." Kelly let go of them and motioned toward the door. "Come meet my new husband. Sharon this is Tom. Tom, Sharon." They shook hands.

"Aggie this is Sgt. Tom Turbulo, Carson City Police Department. Tom, meet Aggie, one of Mike's super-intelligence crew."

"Nice meeting you," said Tom, still frowning. "Afraid I can't ask you in," he said, still blocking the door.

"Of course we can," said Kelly. "Come inside, ladies. We'll get you something to drink."

Tom re-crossed his arms, refusing to move.

"Thomas," said Kelly a bit sharply.

Understanding that tone of voice, Tom stepped aside and allowed the ladies to proceed him into the cottage. Kelly rustled around finding chairs and getting drinks for everyone. Finally she perched on a footstool and clasped her hands between her knees. "Well, maybe you'd better explain," she looked at them.

"Of course," said Sharon. "Mike and Cynthia have been working on your case ever since you called him, Kelly."

"That's right," Agatha agreed. "And so Mike sent us down here to help you out."

"Do whatever we can, or what is needed to keep you safe," Sharon added.

"Don't worry, we are not going to interfere with your blissful honeymoon, on strict orders," said Agatha.

"You won't even know we're here," Sharon promised.

At this, Tom snorted.

Kelly gave him a dirty look. "Thank God you're here," she turned to the ladies. "You can help us investigate these terrorists. At least I think they're terrorists. They must be. I mean, who else would be doing this?"

"Doing what?" chorused Sharon and Agatha.

Tom interrupted. "Really, Kelly, I don't see how these two can be of any help, whatsoever." He waved them away and shook his head.

"Oh my dear, you have no idea what Cynthia's team can do. You name it. They can do it and look innocent in the process."

Tom frowned, "Humph."

"You'll see." Kelly bent her head toward the ladies. "We were just planning how we could snoop into their operation. You can help us." She went on to fill them in.

Later the three women had their heads together, considering a plan. Tom had clearly been overruled.

Chapter 6

CIA Headquarters

Assistant Director Jo Rench opened the daily report from his man in Honduras.

Officer #Q00059667JB
USCIA
Surveillance Report
Subject: Mrs. Kelly Turbulo, American citizen, age 23, Carson City, NM, vacationing Honduras.
Nothing new to report. Mrs. Turbulo and husband, Tom, spent quiet evening in bungalow.
Subject: Ms. Sue Lynn Reese: Suspicious. Claims to be American citizen on leave from office job in San Francisco, plans to attend college. Appears this is false. Reese unable to answer simple questions re: California. Appears to associate with two men. Please check identity of attached picture.

Assistant Director Rench filed the report and dictated a quick answer:

To Officer #Q00059667JB. Continue limited surveillance of the Turbulos. Increase surveillance of Reese and the two male associates. Report any suspicious activity. A check of attached picture reveals no known identity. Referring for forensic evaluation.
JR

Back Alley

The morning was half over. It was already nine o'clock. Su-jin was pushing her men. "Let's get going, people! We should've been out of here a half hour ago."

Two of the men exchanged looks. They covered their mouths and spoke in low tones. "If she hadn't been flirting with the white man at breakfast, we'd've been on time."

"Yeah, right. I agree. She'd better stay off our backs."

"What do you think we should do about it?"

"About the man, you mean?"

"Yeah."

"You think he was white?"

"Maybe. Hard to tell, but I think so, at least partly white. Must have been an American."

"Doesn't matter, she knows we saw them together."

"Yeah, so?"

"We'd better be careful."

"What do you mean? Careful about what?"

"I mean we'd better not turn our backs on her."

"She wouldn't do anything, would she?" he glanced at Su-jin.

The man paused in what he was doing. "Just be careful. That's all I'm going to say." A meaningful look crossed his face.

The second man covered his mouth and nodded in agreement, "I didn't see anything at breakfast, did you?" he winked.

"Not a thing." He bent to continue his work.

Su-jin's people were carefully loading the portable rocket launcher onto the rental trucks. Each piece was wrapped in the new packing materials and then wedged and tied into place. This time there would be no accidents.

Steve

Hidden from view, Steve Spalding watched through powerful binoculars. Just as he suspected, Sue Lynn was not the innocent California office girl that she pretended to be. She had at least six men under her command.

After twenty minutes of hectic activity, the entire crew climbed into vehicles and drove away.

Steve was torn between following them and investigating the operation here. *Well, I can't be everywhere*, he decided. *Something tells me I need to look inside that shipping container.*

Steve left his binoculars, parked his moped behind a Jacaranda tree, donned sunglasses and sauntered down the alley as if he belonged there. Approaching the container he whistled an aimless tune through his teeth, hands in his pockets, scuffing his feet and gazing around. Feeling in his pocket for the exact tool he needed, he checked to make sure the alley was empty. Quickly his hands reached out and snapped open the locked chain with the tool. Three seconds was enough time to disappear inside the container and slide the heavy door closed. Steve paused to listen, every sense on high alert. The container was dark as night. Steve pushed up his sunglasses, moved away from the door and breathed, quietly testing the air and straining for any tiny sound.

Satisfied he was alone and, so far, undetected, Steve flicked on a tiny but powerful flashlight and scanned the space for security alarms or cameras. Then he shined it on a long slender object covered with a large tarp. *Well,*

well, what have we here? Slipping out of his shoes, he crept forward. Lifting up the tarp he gasped and blew out a long breath. *Whoa! Unbelievable!* He spent several minutes inspecting the ballistic missile and taking shots with his cell phone. He hated to use the flash but had to take a chance. Steve took a few long shots and some close-ups of the writing along the side. Carefully he covered the missile with the tarp, trying to remember exactly how it was when he found it. Picking up his shoes he slowly opened the door a crack, just as three mopeds pulled up outside.

Oops, now what th' hell? Steve quickly pulled out his weapon and faded against the wall, straining to hear the voices. He heard women babbling but couldn't make out their words. Then he heard a strong man's voice speaking, "Here, let me have a go at it," in English, followed by scraping sounds and a rattling of the chain. Steve slid into the corner away from the missile, ready to spring. The huge door ground open. A shaft of light shot into the container. "You three wait outside, while I check this out," said the male voice. *Good, only one guy*, thought Steve.

Steve held his breath as a man's foot entered, followed by a well-built body. A few steps into the container, the man let out a whistle. "Kelly, you've got to see this."

"Okay, Tom, help me up."

Kelly? Tom? Good heavens! Steve was flabbergasted. What were these two doing here? Without hesitation, he leaped from the container and broke into a sprint toward his moped. No more than twenty paces later he was hit hard from behind and sprawled flat on his stomach. Hands grasped his arm twisting it behind his back. Steve struggled to get up pushing with his other hand. A knee firmly pushed him down. "Oof!" he exclaimed as pain shot up his arm. He felt a hand in his back pocket.

"Nothing here," said a female voice. "Okay, try the front," said a second. Steve wiggled. "Lay still, mister, if you want to live," threatened the first.

Ah, two women. I can take them. In a sudden move Steve summoned all his strength to flip over. Boom! Down he went, getting a mouthful of dirt.

"Cuff him," said the first.

Steve felt his hands being crossed and cuffs snapped into place. A sharp pain hit his neck as his head was jerked around, a gag stuffed in his mouth and a black cloth covered his eyes. "Up you come," said the second, as the two of them easily lifted him. "Now march." A hand pushed him forward. Steve felt something sharp in the center of his back. Holy crap, two women had him! Steve counted the steps. Twenty steps later, "Up you go. Grab him, Tom." Steve felt more hands lifting him and sliding his body into the container away from the door. "You take him. We'll stand guard." Steve was propped up and shoved against the wall. The door scraped closed. Around the edges of the blindfold, Steve could see a strong light shining in his face.

"Who is this guy?" asked Tom.

"I have no idea," Kelly answered.

"Does he look like the man who followed you?"

"Oh no. I don't think he is one of them."

"But, he must be involved."

"Yeah, probably."

Steve vigorously shook his head.

"Who the hell are you, man?" asked Tom.

Steve mumbled around the gag.

"Let's just leave him in here," suggested Kelly.

Steve shook his head wildly and tried to scream no-no-no. It came out, "Mm-mm-mm."

"All right, all right," said Tom. "Just sit still while I have a look at whatever is in this coffin." The light moved away. Steve started working on his restraints. He fell over onto his side. Back came the light. "Whoever you are, mister, either stay still or I'm going to have to tie your feet too."

"Tom! Look at this," said Kelly.

Obviously, she had discovered the missile. Steve could hear them exclaiming over the thing. He wished they would hurry up before the cavalry came storming back. He strained to listen while they went through the same stages as he had just minutes before. No doubt, they were as stunned as he had been.

"All right, mister whoever you are. We're going to let you talk." The gag was removed. "Now who the hell are you and what are you doing here?"

"Please take off the blindfold," were Steve's first words.

"I don't think so," answered Tom.

"You might as well," said Steve. "I know who you are, Sgt. Tom Turbulo."

Tom remained silent.

"And the woman with you is your wife, Kelly."

Tom and Kelly exchanged looks. Tom pulled in a breath and blew it out.

"All right, mister, you know who *we* are. Who the hell are you?" Tom began rapidly frisking Steve, removing his weapons and phone and handing them to Kelly. "Interesting," said Tom as his examined Steve's special toolkit. "Very sophisticated." He patted and felt inside all Steve's pockets, and thoroughly stroked his body and limbs. "No ID," said Tom. "You're working for someone," he stated. "The only question is whose side are you on."

"I'm on your side," Steve insisted.

"Well, now, we're getting somewhere," said Tom.

"Let me go."

"Can't do that just yet, sir. Sorry."

"Okay, I understand. But, let's get out of here."

"Kelly, have you got enough pictures?" asked Tom.

"Yeah, I'm done," she answered.

"Okay let's cover this thing up."

Steve waited quietly as they flopped the canvas tarp over the missile.

"Before you upload those shots, snap a couple pictures of this guy," said Tom, removing Steve's blindfold.

"Hold still, mister," said Kelly aiming her cell phone at Steve. "Now, say cheese."

Steve stared at the phone, stone-faced.

"You go first and get those pictures uploaded, honey," said Tom. Kelly shoved open the door and hopped down. Tom moved over to Steve. "Okay, let's get you up," he said. "Can you roll over onto your knees?"

"Yeah, with a little help," Steve groaned. He turned part-way over, while Tom grabbed him under the armpits. Together they managed to get Steve onto his feet and over to the door.

"My shoes," said Steve.

Tom shined his light around. "Here you go. Put these on."

Steve slipped awkwardly into the shoes. "Shall I jump?"

"Why don't you just sit down and slide off onto your feet?" said Tom. "Sharon, Aggie, can you give us a hand here?"

Steve looked up to see Kelly off to the side, now working her cell phone as two beautiful women approached him. They each took an arm and helped him down.

"You!" Steve exclaimed.

They laughed. "Yup, we took you down, didn't we?" Agatha mocked.

"We're stronger than we look," Sharon laughed. "Come along, mister, you're going with us." She took his arm and hustled him onto the passenger seat. She hopped in front, grabbed the handles and placed one foot on the starter. "Off we go."

"No, no, wait a minute," Steve said. "I'm still handcuffed."

"Too bad for you."

"But … my moped."

"Oh yeah? Well where did you hide it?"

"Up behind that Jacaranda."

"Ya' hear that, Aggie?" asked Sharon. Agatha took off jogging toward the tree. Sharon followed with her passenger balanced precariously on the back. "Hang on with your knees," she yelled. Hands cuffed behind his back, bouncing around, Steve practically had his legs wrapped around her.

Kelly mounted her bike and waited for Tom.

Tom closed the shipping container, restored the chain and snapped the lock closed. "Go on, I'll catch up," he said.

Villa

"Nice place you have here," Steve remarked, flexing his arms and rubbing his hands where the cuffs had chaffed him.

Sharon pocketed the cuffs. "Be good, mister, or I'll tie you up."

"Absolutely," Steve promised with a grin. "Can't cross a beautiful woman."

Kelly handed him a tall glass of iced tea. "Thanks," he said and saluted her with the glass before taking a big gulp.

Kelly retrieved two more drinks from a tray and handed them to Sharon and Agatha, who nodded their thanks and drank. "This hits the spot," said Sharon as she leaned back on the sofa next to Agatha.

Kelly took a side chair and placed her drink on a small table. 'Well, what do you think?" she asked the group. "Should I turn on the air?"

Agatha looked at Sharon who wiped beads of sweat from her brow. "D.C. is hot but nothing like this," she agreed. "But, let's just open up the place and enjoy the ocean breeze."

Kelly moved to open the sliding doors on both ends of the room.

"Here, let me help," said Steve, jumping up. Sharon and Agatha leaned forward ready to stop him if he bolted.

"Thanks, you can get that one," Kelly pointed.

"How about that window?" Steve gestured toward the kitchen.

"Sure, that'll be good," answered Kelly

Steve unlocked the kitchen window and cranked it open. "How's that?" he asked.

"Perfect," said Sharon, lifting her arms and fanning her pits with her blouse. "Ah." She smiled and turned her head to catch the breeze.

The four people settled back down and resumed nursing their iced tea. One drink remained on the tray.

"Tom should be here any minute, right?" asked Sharon.

Kelly wrinkled her brow. "Um, he said he'd catch up."

"Was he going somewhere else?"

"All he said was, 'You go on, I'll catch up'."

"Oh."

The three women sat quietly, thinking for a minute. Then they suddenly turned on Steve. "You weren't alone were you?" Sharon spoke for them. "You son-of-a-bitch!" She leaped from her seat toward him.

"No-no-no!" Steve threw up his hands. "It's just me. Honest to God!"

She launched toward his throat. Steve fended her off, grabbing her arms and pinning them to her sides.

"Woof!" said Sharon falling into his lap. She struggled to get loose.

Steve grinned showing perfect white teeth against his tanned skin. "Hold on, lady. Hold on just a minute. I can explain."

Sharon lifted her head, inches from his face. "Go ahead. Explain."

"Will you stop fighting?" he bargained.

Kelly and Agatha were laughing, enjoying the show.

Sharon nodded, "Maybe."

"Okay," Steve loosened his grip. Sharon leaned back, still in his lap, her body against his chest. Steve gazed into her eyes and gulped. He cleared his throat. "Uhm, well … ah."

Sharon half rose, propping up with her elbow. She frowned and squinted into his face. "Go on, you jackass, before I take you apart with my bare hands and teeth."

Steve laughed aloud. He had a sudden urge to tickle her but thought better of it. "Honestly, sweetheart, I'm on your side."

"How do you know what side I'm on, jerk?" she tossed.

"Not sure. I just hope and pray it's my side." He looked around at the three. "Otherwise, I'm in deep doo-doo." He grinned sheepishly, looking boyishly handsome with his hair mussed, his skin a healthy golden-bronze tan.

"Can I get up now," Sharon asked, seeming to simmer down.

Steve shoved her cute little butt. "Upsy-daisy."

Sharon smoothed away imaginary wrinkles, gave him a sidewise frown and moved back toward her seat remaining standing with hands on hips. "I still demand to know what you did with Tom," she crossed her arms and tried to keep her voice even.

"Honey believe me I did nothing with Tom. He's gone off somewhere on his own. I work alone, dammit. I wish to hell I had some help." He raised his hands in mock surrender. "I'm going crazy, working all alone, trying to keep track of Kelly, here, and watch out for those stupid terrorists or whoever the heck they are. I mean what are we going to do about them, for God's sake?"

"Look's like we have a shared mission," Agatha stated, calmly.

She and Kelly looked at each other.

"Ya' think?" asked Kelly.

"Well, seems to me that mister big-shot here isn't going to tell us whose he is or why he was keeping track of Kelly. My sense is he's undercover." She looked at Steve who didn't blink.

"Why do you say that?" Kelly wondered aloud.

"Well, he's been spying on you, Kelly. Seems to me he could have harmed you if he wanted to. Maybe that's not his purpose." She shrugged. "Also, I'll bet a lot of money that Tom is still snooping around down there."

Kelly wrung her hands, "I'm really worried. Those men are vicious. They could be anywhere, you know." She looked anxiously at her watch. "The gift shop probably opens up soon. Maybe we should go back down there."

"How many men were there?" Steve interrupted.

Kelly turned toward him. "Well, there were maybe six, I think, who came after me. Could be more of them hiding somewhere, you know."

"Well, let me help put your mind at ease, Kelly," said Steve, kindly. "I can tell you this much. I was watching them, before you rode up. There were six men with a woman. She was giving orders like a boss or a commander. I watched them load some heavy equipment from the shipping container into a small truck and a van."

Sharon finally took her seat and stared, fascinated, at Steve. The other two women leaned forward. "Go on," said Kelly.

"They all got into the vehicles and drove away."

"Did you have any idea where they were going?"

"No I didn't. But, I'm sure that Tom will hear them if they return. He won't stick around."

Kelly unclasped her hands, blew out a breath and leaned back.

"You can relax for now," Steve looked at her sympathetically. "He'll be back soon."

"Yeah," said Agatha. "I think we should wait here. I know you're worried, sweetie, but look at it this way. You married a cop."

Kelly half-smiled. "Yeah, I sure did," she moaned.

Steve looked at his watch. "If he's not back in thirty minutes, I'll go after him," he offered.

"All right," said Kelly, "twenty minutes and that's all."

Steve nodded.

"I don't think so," said Sharon. "You're staying right here, mister until we figure out who you are. I'll go, Kelly."

Just then, Kelly's phone beeped. "Incoming message," she said, picking up the phone. "It's from Cynthia." She fingered the message and read aloud. "Missile is intermediate range. Markings are Korean. The hunk is CIA. Goes by Steve." Wide-eyed, they all stared at Steve.

A lopsided grin crossed his mouth. "Um, busted," he said.

Agatha broke the silence, "No surprise. So, Mr. Steve-Um-Busted, why does the CIA care about Kelly, anyway?"

"Always keeping my eyes open, I ran into that suspicious group. But, my primary orders were to make sure nothing happened to Kelly Turbulo. I guess someone important was worried about her."

"No kidding? Me, too," said Agatha, laughing. "Looks like you've got some high-powered protection, lady."

"I don't need protection, thank you very much," Kelly huffed. "I've got my own personal cop. You people need to get lost," she glanced at her watch.

"Not so fast," said Sharon snapping out of her reverie over the handsome CIA agent. "We're here now and you've been snooping in some very dangerous places."

"We're here to do a job, Kelly," said Agatha, "and we aren't leaving anytime soon."

Steve patted his pockets. "Would you mind giving me back my cell phone? I need to report in."

The three women looked at each other and shook their heads. "We don't have it," said Sharon.

"Probably Tom has it," said Kelly, "so I guess you're just going to have to wait for him."

Chapter 7

VFW Convention

The wheels on Air Force One squawked as it touched down in Kansas City carrying US President Gerard Bigelow and his entourage of aides and reporters. A waiting helicopter whisked him to the convention center where a few thousand military veterans packed the vast room. Large screens outside projected the proceedings for the thousands more who would not fit inside. Cameras from all the major networks and local TV stations were crowded across the rear of the auditorium. Back in the studios of the various networks talking heads continued reading from their teleprompters, filling in while the networks waited for the president to arrive. A parade of politicians running for office kept the audience busy building the anticipation.

"Hail to the Chief" started playing. A cry went up from the people closest to the stage entrance. Wild cheering swelled as the crowd realized who had entered. Cameras swung onto the president and his wife, First Lady Janette Bigelow, as they entered hand in hand. Gerard smiled broadly waving at the crowd, strolling across the stage and pointing from time to time.

"USA! USA! USA!" screamed the crowd.

During a full three minutes of cheering, the studio announcers criticized Janette's outfit. "Look at that ugly pants suit."

"I hear it was made by some obscure French designer."

"Do you think she's gained weight?"

85

"What's that on her feet?"

"Are those shoes?"

"Cowboy boots!"

"Ugh, they're better suited to the barnyard."

"Or riding the range."

"Works well on a rancher's wife."

"Yeah, he's a cowboy, all right."

The sound switched instantly as Gerard Bigelow stepped to the mike. "Thank you, thank you." The applause died down. "Hello, Kansas!" he exclaimed, followed by another full minute of cheering. He waved both arms, grinned, paced, and nodded.

After thanking his hosts, recognizing the politicians present, one by one, and calling on an aging WWII vet, Bigelow went through his standard campaign speech, pausing at all the right moments. The audience knew when to cheer and the president was an expert at working the crowd. "I'm pleased to announce the largest military budget in history, seven hundred billion dollars. We have the greatest military force in the history of the planet." The audience roared its approval.

Meanwhile, across the oceans, on the other side of the planet, Dear Leader and General Lee listened to Bigelow's speech. "Not for long," DL commented. "Soon the tiny mouse is going to take down the over-bloated elephant." General Lee chuckled and nodded agreement. They watched a small display screen as Bigelow continued his show.

"Would each service branch please stand and be acknowledged?" asked Bigelow. "First the Army veterans. Thank you for your service. Thank you Army veterans… now the Navy … Air Force … Border Patrol." Bigelow paused to clap after each one, leading the cheering. "Coast Guard … Space Force … do we have any members of our new Space Force? Let's have a big hand for all our veterans." He applauded. "Thank you very much. You may be seated."

The announcement of a newly formed space force went unnoticed everywhere except on the other side of the world. DL exclaimed, "What did he just say?"

"Uh," answered Lee. "Maybe he was kidding. You know how Bigelow likes to brag and crack jokes."

DL stroked his chin and frowned. "American Space Force," he muttered. "Probably sitting at computers, fending off cyber-attacks," he chuckled. "Those hands will all melt to their keyboards when we unleash Jagjeon Haeg Gyeoul, Operation Nuclear Winter." Turning to Lee, he said, "Switch that moron off, and tell me how our scientists are doing. Let's talk about progress with JHG."

"The satellite design is completed," said Lee as he reached to shut off the video. "We have two miniature satellites that will mount on the missiles."

"Good. We need to get them shipped to Honduras," said DL.

"We are taking care of that."

"And the bomb?"

Lee bit his lip. "There are still a few things to work out."

"What things?" demanded DL.

"This is extremely complicated."

"So?"

"Well, you have the problem of the weight limitations, the expulsion device, the guidance computers, the remote controller all needing to fit under the weight limit. And then the bomb..." his voice trailed off.

DL slammed his fist on the desk. "Nonsense!" he screamed. "You aren't doing your job."

Lee jumped and started to rise. Then he slouched and bowed his head. "Yes, sir," he mumbled.

"Go out there and teach those lazy people some discipline. Assemble them in the yard. Make an example of a few. That will hurry them along."

"Yes, sir."

"Now, get out of my sight!" screamed DL.

Remaining bowed, General Lee crept out of the office.

Washington D.C. Sen. McBride's Office

Cynthia Patterson-Eastman drummed her red fingernails on the desk. A tiny frown etched her otherwise satin-skin. *Those pesky CIA people*, she mused. *Putting on that innocent act, all the while spying on Kelly and Tom. Well, we'll just see about that.* She picked up the phone and hit the direct-dial number. "Assistant Director Rench, please. Senator McBride's office calling."

"Hello?"

"Jo, this is Cynthia Patterson."

"Cynthia! How nice of you to call! I hope this isn't business." CIA Assistant Director Jo Rench was a bit of a harmless flirt.

"Sorry, Jo, I'm on a mission."

"Shucks. I thought I was going to get lucky."

Cynthia laughed and went along. "Heh, you're the luckiest guy I know. Too bad it didn't rub off on your operative in Honduras."

"Don't tell me," said Jo. "I don't need any more bad news."

"Not too bad, Jo. Steve's fine and everything's cool. It's just that my operatives took him down, disarmed and cuffed him." She laughed aloud. "It was a simple matter to ID your man. Tsk-tsk."

Jo groaned.

"Yup. Of course we're not rubbing it in, you know."

"Oh no, you wouldn't do that, would you?"

Cynthia chuckled. "All we require is a little cooperation to let him go."

"Blackmail!" exclaimed Jo. "You can't do this."

"Not at all, Jo," Cynthia soothed. "This is entirely voluntary. For our part, we wouldn't harm a hair on Steve's handsome head. Our girls are enjoying his company."

"All right," Jo sighed. "Name your price."

"Cooperation."

"Is that all?"

"Transparency."

"Can't do that."

"Sure you can. You're the master leaker."

"Wash your mouth out!"

Cynthia laughed. "In return we'll let you in on our interesting discoveries concerning a certain terrorist cell in Honduras hiding a ballistic missile capable of reaching the United States."

"It's your duty to report it, young lady," Jo retorted.

"Sure, sure," said Cynthia. "We'll be more than happy to work with your guy as a team. Share and share alike."

"Why do I need you, when I already have a good man in Honduras?" Jo asked, weakly.

"Your man seems to have misplaced his top-secret cell phone, his weapon, ID and special tool-kit."

"Good grief, woman!" Jo moaned. "Do I have to go down there, myself?"

"Not yet. Just send fresh orders to Steve to cooperate with my team. Easy-peezy."

"No such thing as easy. I need a vacation."

"So? Don't we all?"

"All right, you win. But, you owe me dinner."

"Thanks, Jo."

Assistant Director Rench dictated a quick communique:

To Officer #Q00059667JB.
Continue surveillance of subject.
Join forces with McBride operatives and CCPD officer.
Report explanation of capture, immediately.
JR

Cynthia buzzed her boss, "Can I see you a minute?"
"Sure, come on in."
Cynthia walked into Senator McBride's office.
Mike looked up from the papers on his desk. "Hi, Ms. Patterson, how's it going?"
"Quite well, actually," Cynthia replied. "I heard from our team in Honduras. They met up with Kelly and Tom."
"I'll bet they were a little surprised."
"P.O.'d might be a better description. Well, no, actually Kelly was gracious. Tom, not so much. But, he got over it. The four of them teamed up and went over to check out the gift shop where Kelly was assaulted."
"The Pigsty?"
"That's it."
"Did they go inside?"
"Not yet. But, they found a couple interesting things inside the shipping container parked out back."
"Oh yeah?"
Cynthia went on to show Mike the rocket pictures and fill him in on the encounter with the CIA agent.
"That sneaky Jo Rench!" said Mike with a wry grin.
"His boy was no match for our team," Cynthia opined.

The Villa

Several anxious minutes passed before Tom's moped pulled up outside the villa. Kelly ran to the door.

Tom parked the bike and dismounted as Kelly ran into his arms. "Thank God you're here, darling, I was so worried."

"Oh, I'm sorry sweetheart. I was fine."

"What were you doing?" she countered as they walked arm in arm to the front door.

"I was checking around inside that secret computer lab."

"How did you get in?"

In answer, Tom showed her the tool he used. "I had this handy little gadget."

"How did you pass that through airport security?"

"I didn't. I just picked this out of the guy's pocket when we caught him in the shipping container. I think I'll keep this thing. Works like a charm."

"Cool."

Tom opened the door for her to go inside. "Hi, everybody." He moved into Kelly's seat.

"Here, honey," said Kelly, handing him the last glass of iced tea. "I'm afraid it's gotten warm. Let me get you some more ice."

"No, this is fine, just so it's still wet," he grinned and glanced around. "Oh sorry, I took your chair, didn't I?"

"That's okay. I'll sit here," said Kelly, moving the tray aside and perching beside him on a small table.

"Well, well," said Tom, squinting one eye at Steve, "you ladies work fast."

"Fast and deadly," amended Agatha.

"Sergeant Tom Turbulo, meet Special Agent Steve Somebody," said Sharon. "By the way, what's your name, mister?" she asked.

The two men eyed each other.

"FBI?" asked Tom.

Steve merely stared.

"He doesn't talk much, does he?" asked Tom.

Sharon explained, "Poor little boy, he's lost without his cell phone."

Tom reached into his pocket and pulled out the phone. "What have we here?" he mused. "Let's see what it can tell us."

"Won't work without the password," Steve groused.

Tom gave him a look and raised one eyebrow. "First, let's see if it's charged." Tom held the button and watched while the phone turned on. "Give me your thumb, Steve."

Steve immediately sat on his hands.

Tom held the phone up to Steve's face. It came online and beeped three times.

"Mm, let's see who sent you a message," said Tom. He tapped the phone, waited a second and tapped again. "Here we go, special agent Steve No-name. Ah, so you are officer #Q00059667JB. You have a message from someone. It's even signed JR. Who could that be? Hm?"

"Come on, Tom, read the damned letter," said Sharon.

"Oh, okay. It says, continue surveillance of subject. Join forces with McBride operatives and CCPD officer. Report explanation of capture, immediately."

The girls broke out laughing at Steve.

"Ha-ha. Gotcha!" said Sharon.

"Word travels," observed Agatha. "Wonder how JR figured that out so fast."

"I told Cynthia," offered Kelly.

"Mm," Agatha nodded toward Steve. "Looks like we've got a new partner."

"Yup," said Sharon, 'but I don't think we can trust him, do you?"

"Good question," said Agatha. "What do you think, Tom?"

Tom was busy looking back in the history on Steve's phone. "Well, I can see that this sender of past messages is named 'Jo' if that tells you anything."

"That would be Jo Rench, assistant CIA director in charge of Central America," said Sharon.

"That's good enough for me," said Tom. "Welcome to the team, Steve."

"Thanks, I think," said Steve. "Can I have my phone, now?"

"Hold your horses," said Tom scrolling through the messages. Just then the phone emitted three quick beeps. "Hm, a new message."

Steve started to rise from his chair. Sharon and Agatha were on him instantly. "Sit still, mister."

"Ah, you have another message from your boss," Tom reported. "It says 'Forensics shows subject is Asian with facial plastic surgery. So far, unrecognizable. Will order bone reconstruction, if desired.' It's signed Jo."

"Give me that thing," Steve pleaded.

"Hold on," said Tom. He typed a reply to Jo, reading aloud, "Clearly Asian woman subject is unfriendly. Will continue to investigate. No need here for bone reconstruction. Your decision. Signed, Sgt. Thomas Turbulo, CCPD. PS. Steve will cooperate."

"Dammit, Turbulo!" said Steve.

Tom laughed. "Shouldn't I hit Send?"

"Oh hell, what difference does it make? My career is shot, anyway."

Tom looked at the women. "What do you say, girls? Should I send this?"

"Let's wait on that," said Sharon. "Maybe we can rescue this poor slob."

Tom hit delete and handed the phone to Steve, with a grin. "Here you go, pal. The ladies have spoken. You're back in business."

"Thanks."

"What's a bone reconstruction?" asked Kelly.

Tom answered, "It's a rather slow and costly procedure used by a forensic sculptor to reconstruct a face from the bones. Usually it is used to identify a skull. Not sure how they do it from a picture."

"Wow," said Kelly.

"A skilled forensic sculptor can make a cast from the skull and then make a plaster head and add all the features, skin, hair, eyes. It's an amazing thing to see," said Agatha.

"Yeah, I've seen it done," Sharon added, "but I think Jo may be referring to something different, a computer process whereby they construct a 3-D image from a two-dimensional picture. From that they can postulate what the original face would look like without the plastic surgery. There is a certain amount of guesswork involved."

"In that case, there's no harm in trying, I suppose," said Kelly. "But it sounds like they may have done that already. Didn't Jo say forensics shows the woman is Asian?"

Tom nodded, "Yeah, he did. So, in the meantime, I think we can learn more about what's going on right here in Honduras."

"So, tell us what you were up to," said Kelly, leaning forward toward Tom. "I was so worried."

"I'm sorry about that, honey, but I was very careful." He gazed earnestly into her eyes.

"Good," said Kelly. "I just wish I had known what you were doing."

"Would that have eased your mind?"

"I don't know. Maybe. At least I would have only one thing to worry about instead of thinking it could have been any number of horrible things. The girls thought that Steve had an accomplice who had nabbed you."

"Oh, yeah, I see your point." He leaned forward and took her hand. "Well, we two have got a lot to learn, honey, and our whole lives to work on it. I'll do my best."

"So will I, dearest." She looked at him lovingly and squeezed his hand.

Sharon cleared her throat. "Uh, you were saying, Tom."

Tom leaned back. "Oh yeah. Well, my plan was to poke around in their office and see if I could learn anything about that rocket we found. But first I booby-trapped the two entrances so I would hear if someone was coming. That way I would have a couple seconds warning."

"How did you do that?" asked Kelly.

"Well, there are lots of ways, but this time I had to use what was handy. I found something in the gift shop that worked. I'll tell you about that later. What I wanted to see was what sort of plan they have to use that missile. From what I could tell it wasn't a solid-state missile, I mean it would require a launch platform. I certainly didn't see one and I'm sure the US would know if there was one around here anywhere."

Steve interrupted. "They have one."

"What?"

"I said they have one. I think the terrorists have a launch platform."

"You saw one?"

"No, but I saw them loading enough large pieces onto a truck this morning that could have been parts to a portable launch platform."

"Steve had been spying on them, before we drove up," explained Sharon.

"Yeah, as I was telling the ladies before you got back, I watched them for maybe half an hour, loading heavy pieces into a truck and a van. Couldn't tell what the pieces were as they were all wrapped up. Then the seven people drove off in the vehicles, giving me the opportunity to look around inside the shipping container. That's when you all pulled up and threw a monkey wrench into my day."

"Change of plans, right?" Agatha commented.

"You sure screwed things up, thank you very much."

"No problem. Happy to oblige."

Steve made a face and rolled his eyes, "Yeah right," he said, sarcastically.

"Well, anyway," he continued. "I'm guessing, now, those pieces were part of a portable launch platform. If so, we ought to be able to find it with our CIA satellite. So, Tom, did you learn anything during your look-around?"

"Well, basically it was just a computer lab, like Kelly thought. They have display-maps of the United States and Canada. We are going to have to get into their computers to find out more. I looked around for any paper files and didn't see any. They must keep everything on computer."

"Did you get into their computers?"

"No."

"We'll have to go back," Steve remarked.

"Maybe this is a dumb question, but, why would they have just one rocket, without any warhead?" asked Kelly.

"Why go to all that trouble for one unmanned intermediate range missile?" agreed Agatha.

"Yup. Good question," Sharon nodded.

Tom had an idea of the answer to that question, which he was keeping to himself for now. It was too awful to contemplate. "Oh, Kelly, there was one thing that would interest you," he changed the subject.

"Oh yeah, what was that?"

"I saw a backpack that looked exactly like your old one. It even had the initials KM."

"What color was it?"

"Blue."

"Yeah, it must be the one I lost when that goon attacked me. I'd forgotten that part. After I escaped from his arms and ran like crazy, he caught up and grabbed me by my backpack. I let go of the pack and got away."

"I looked through the pack and didn't see any identification," said Tom.

"No I had that hidden in my fanny pack."

"Smart move Kelly."

"Yeah my mama taught me well—where to hide the good stuff," she grinned.

Steve was relaxed, now, working on his cell phone, probably typing some sort of excuse to JR. The rest of the team went on with planning their next move. Obviously they had to get into the computer lab, but they needed to do it safely. The discussion continued. How could they keep track of the terrorist's whereabouts, with a tracking device maybe? Tom wanted to charge down there right now. Sharon and Aggie favored getting help from Cynthia. Kelly urged caution.

Steve suggested they program each other's numbers into their cell phones. "If we are going to be working together, we need to be able to get in touch and stay in touch, don't you think?" he asked.

"Absolutely," said Tom as he pulled out his phone. The others followed. "I'm going to make it easy by putting all of you on one button." They all agreed and started typing.

They paused as Steve's phone signaled a message. Steve read the message and then looked up. "No worries, people. We've called out the big guns. The Air Force is sending a reconnaissance plane. Also, the Space Force is directing a spy satellite pass-over."

"Did you say Space Force?" asked Tom, incredulously. "What's that?"

"Yeah, I said Space Force. You didn't know? Where have you been, mate?"

"Never heard of it," said Kelly, defending her husband. She looked at Sharon and Agatha. They both shrugged.

"Welcome to the 21st century," Steve snarled, apparently pleased to recover some of his macho. "We'll get a report later today. In the meantime, I have a couple things to do." He rose to leave.

"Not so fast," said Sharon. "How do we know you'll be back?"

"You don't, do you?" he gave her a sly look, daring her to stop him.

"I'm coming with you." Sharon rose and dangled the moped keys in his face. "You won't get far without these."

Steve broke out in a smile. "Okay, baby, let's go."

Sharon and Steve

They left together. Sharon handed Steve his helmet and fastened her own in place.

Steve started to climb onto the driver's seat. "Oh no you don't," snapped Sharon. "Get in the back, buster." She pointed with a thumb.

Steve gave her his best movie-star-Clark-Gable-bad-guy look, "My pleasure, sweetheart." He mounted the backseat and gestured to her. "Be my guest."

Sharon climbed on the front, put the key in the lock and turned it on. Steve wrapped his arms and knees around her and snuggled his body close to her backside. Sharon pretended not to notice, while trying to ignore the electric tingling sensation invading her lower body. She gulped and politely asked, "Where to?"

Steve's lower parts responded uncomfortably. He adjusted his seat and cleared his throat, "Uh, well, first, let's just swing by The Pigsty to see if the people are back, and then we'll decide what to do. I may have you drive me to The Blue Roof Resort. That's where the woman is staying."

Sharon nodded and started the motor. Off they went with the wind blowing their hair.

They found the van had returned and was parked in the alley behind the gift shop. Sharon cruised by slowly and then stopped one block over. "What do you think?" she asked.

"Could be they have returned and taken the truck back to the rental agency. I don't think it's safe to go in. Probably someone is there running the store."

"Yeah, I see what you mean."

"We need to put a tracking device on that van."

"Good idea."

"Do you have one handy?"

Sharon laughed, "What do you think?"

Steve laughed and moved his hands over her body, just grazing her breasts. "Hm, I don't feel anything," he purred.

"Oh yeah? Seems to me you're feeling plenty. Now, stop it," she wiggled and giggled, without sounding convincing. He moved his hand again. "I said stop that Mr.

99

Steve Somebody!" she said with a bit more force and slapped his hand. "I don't even know you."

"Patience, baby," he breathed on her neck. "I plan to get better acquainted."

Sharon shook her head and blew out a breath. "Let's stick to business, shall we?"

"Okay, okay. Business. Let's see. Where were we? Oh yeah, the tracking device. Have you got one handy?"

"You know damn well, I don't," she grumbled.

"Well I have two options to offer. We could swing by my hotel and pick one up, or we could take the one off of here and put it on the van."

"What!" You put a tracking device on my moped! Dammit, Steve."

"Like Tom said, hold your horses, baby. I did no such thing."

"Oh. Well, then, who did?"

"The rental agencies almost always do, to prevent theft."

"So, they probably have one on the van already," said Sharon, acting a bit chagrined for having popped off at him.

"Probably. Unless those guys are smart enough to remove it."

"Okay, let's assume they aren't and the tracking device is still in place. How do we tap into it?"

"We'd have to have the tracking number," Steve answered.

"And I suppose you smart CIA agents know exactly how to get that," she said with a note of sarcasm.

"We have our ways," said Steve, neither confirming nor denying. "It would be a lot easier to use the one off the moped. I could figure it out from the paperwork and the vin number. But, it would be safer to get one from my supply."

"Why safer?"

"You want to know everything, don't you, sweetheart?"

"Yup."

"Well the CIA equipment will be the best quality. It will have a great battery, be sturdier, more reliable and will broadcast longer and farther. Also, keeping your bike wired is simply a good safety factor. The one thing I will do, with your permission, is to program the tracking number into my cell phone, just in case anything happens to you, I can find you. Okay?"

"Why, Steve, I didn't know you cared," she flirted flippantly, in an attempt to cover her real feelings.

"Believe me, my dear, I'm just as surprised as you are."

Their eyes met and held.

Sharon dropped Steve off at his scooter bike and she went back to her hotel room. Steve was able to do some things on his own. He checked in with his informer at the Blue Roof Resort and determined that Sue Lynn was in her room, meaning her gang was obviously back in town. He made plans for someone to follow her so he would know the next time she left town with her people. From his supply, he took a couple tracking devices to attach on the van. As luck would have it he found a luxury car parked next to the van, so he was able to put devices on both vehicles. Now he would know if they left the area and would be able to track them. Hopefully the devices would disclose their secret launching site and let him know when the computer lab was unprotected long enough to allow him inside.

101

After getting their solemn promise that they would not leave the immediate area of their resort, Agatha decided it was safe to leave the honeymooners alone for now. She returned to their hotel room. "Hi Sharon," she greeted her roommate. "Any news?"

"Nothing urgent," said Sharon. "Why don't we sit out on the deck and I'll bring you up to date? Go ahead, I'll get a couple beers."

"Tea is good for me, thanks," said Agatha.

Settled on the deck overlooking the ocean, the two filled each other in on the news.

"I left Steve with his motor-scooter," Sharon began. "I think he was going to put a tracking device on the van we found parked behind the gift shop, and then he was going to do some snooping around. That's about all I know."

"Same here. Not much happening until we get the results of the flyover. How about you and Steve?"

"What do you mean, me and Steve?"

"I mean how about the sparks I saw flying between the two of you?"

"No sparks flying," Sharon denied, looking away.

"Um-hm, no sparks, yeah right." Agatha laughed.

"Nonsense. There is no me and Steve."

"You gotta admit, he's a hunk."

"Hunk-schmunk. The day you see me take up with a CIA agent is the day frost forms in hell. Have you heard anything from Cynthia?"

"Not recently. You're changing the subject."

"Yes, I am. Maybe we'd better check in with Cynthia."

Washington DC

Back in Washington, Cynthia was having a frustrating day getting absolutely nowhere trying to deal with the bureaucratic establishment. Senator McBride was off somewhere working on committee responsibilities. Assistant Director Jo Rench didn't answer the phone and none of her other contacts in the CIA knew anything. Correct that. Either they knew nothing or they were not saying. Whatever the reason, CIA did not seem to care or think Kelly McBride-Turbulo's issue was the least bit important. Cynthia was getting the brush-off and she did not like it.

Island Boredom

On the island, the hardest part of any operation was the waiting. Everyone tried to deal with it in his or her own way. For the honeymooners, waiting was not a problem. They knew exactly how to entertain themselves.

Steve roamed the restaurants and beaches, acting the playboy role and keeping his eyes open.

Sharon and Agatha were at loose ends. They walked the beach, sought out a couple of different restaurants and bars, wandered the lobby and tried the game room. Finally they each picked up a novel from the sparse paperback lending library, slathered themselves with suntan lotion, commandeered two lounge chairs near the pool and settled down for some reading time interspersed with naps.

It was dark outside by the time Steve messaged everyone to meet at the villa. The five gathered in the small living area.

"Well, folks, I heard from headquarters," Steve began. "The flyover and the satellite imaging turned up exactly nothing. No launch sites, no new buildings, machines or construction areas that weren't there three months ago."

"Strange," said Tom.

"Yeah, it is. But if anything was there, they would have found it."

"You're sure?" asked Sharon.

Steve gave her an incredulous look. "Yes, of course I'm sure. Those cameras can see a hair on a wart on a frog on a bump on a log in a hole in the bottom of the sea, don'cha know?"

Sharon laughed and sang out, "There's a hole...there's a hole...there's a hole in the bottom of the sea," clapping her hands and rolling around in near hysterics.

Tom and Kelly looked at her, amazed. "What have you been drinking?" asked Kelly.

"Mai-tais," laughed Sharon licking her lips. "Oops," she covered her mouth. "Sorry."

Steve sang, "There's a log in the hole in the bottom of the sea."

Agatha sang, "There's a bump on a log in a hole in the bottom of the sea."

Steve and Sharon joined arms and sang out, "There's a hole...there's a hole...there's a hole in the bottom of the sea."

Tom and Kelly frowned at them, puzzled.

Sharon, Agatha and Steve guffawed and pointed at the Turbulos, while holding their sides. "Ya' wanna hear the rest?" Sharon gasped.

"Um," Tom shrugged, "Is there any way to stop you?"

Sharon shook her head and belted out, "There's a frog on a bump on a log in a hole in the bottom of the sea, there's a frog on a bump on a log in a hole in the bottom of the sea," Agatha and Steve joined in, "There's a hole, there's a hole, there's a hole in the bottom of the sea." And then they laughed. "Steve you're next."

Steve sang, "There's a wart on the frog on the bump on a log in a hole in the bottom of the sea." He gasped for breath. "There's a wart on the frog on the bump on a log in a hole in the bottom of the sea." Tom, Sharon and Agatha joined in, "There's a hole, there's a hole, there's a hole in the bottom of the sea."

Meanwhile, Kelly was uncapping some beers. She walked in with a tray of beers and passed them around. "Drink up, everyone."

Each one took a beer and clicked them together. "To our success," said Kelly. "To long life," said Sharon. "To happiness," said Tom. "To good sex," said Steve. "Cheers," said Agatha, frowning at Steve. "Now, all together: There's a hole in the bottom of the sea. There's a hole in the bottom of the sea. There's a hair on a wart on a frog on a bump on a log on a hole in the bottom of the sea... There's a hair on a wart on a frog on a bump on a log on a hole in the bottom of the sea.. Everybody! There's a hole, there's a hoh—oh---oh-olll! There's a hole in the bottom, the very bottom of the deep, deep bottom of ... the ...se-e-e-ah-ah-ee."

Laughing wildly Steve pronounced, "I'll drink to that." Everyone cheered and tipped up their beers.

Twenty minutes later, Sharon announced, "Thanks a lot, Kelly but we need to push off." She stood.

"I'd better see you home," said Steve setting his empty beer bottle on a nearby table and rising toward the door.

"No need," said Sharon. "We're just up the beach."

"Did you walk?" asked Kelly.

"Yes, it's the next building over."

"Good night," said Agatha, joining Sharon and Steve.

Kelly and Tom moved to see their guests out. Tom kissed the ladies' cheeks. Steve kissed Kelly and shook Tom's hand.

"Listen, Tom, I'll call you in the morning if it looks like the terrorists are leaving."

"Sounds like a plan," said Tom. "Not too early."

"Won't be before nine," he laughed, "I think."

"Perfect," said Tom opening the door. "Here's your hat, what's your hurry?"

Steve chortled, "I can take a hint," put one arm around Sharon and left by the door.

Chapter 8

Next Morning

The sun peeking around the drapes in the bedroom window failed to arouse Tom from his dreams.

"Wake up, sleepyhead," called his wife's cheerful voice as she entered the bedroom carrying a breakfast tray. She moved to open the drapes and pull back the curtains.

Tom groaned and rolled over.

Kelly propped up his pillows and gave his butt a playful smack. "Up you go, darling." She helped him onto the pillows. "Coffee?"

Tom mumbled and rubbed his eyes open. He reached out his hand and Kelly placed a half cup of coffee in it.

"Here's your coffee, dear. Careful now."

Tom blew on it and took a test sip, then another. He looked at her, realizing now where he was. He took a long sip. "Thank you."

"Good morning, darling."

"Good morning, sweetheart." He took another sip. "I could get used to this." He grinned.

"Absolutely, it's just part of the service," she poured her own cup and took a sip. "I brought some juice, as well. Freshly squeezed."

"Thank you, that's great."

"I thought I'd see what you like for breakfast, eggs and toast?"

"I'm not hungry just yet."

"Okay then, would you like the island newspaper, cell phone or your computer with your coffee?"

"All three, for now. Let me enjoy this for a few minutes and then we can have breakfast on the patio. Okay?"

"Sounds perfect."

"By the way, last night was great. Was it okay for you?"

"Yes, fantastic," she replied.

"You're not too sore?"

"Just a little. I'll be okay. Don't worry about it."

"Okay. You'll let me know, won't you?"

"I will," she smiled.

"You're wonderful," he said, gazing straight at her.

"I'll be on the patio when you're ready for breakfast." She filled his coffee cup, topped off her own and left. "Enjoy."

Senator McBride's Offices

Yesterday had been such a drag, Cynthia was hoping for more progress today.

Her private computer chirped. It was a Skype call from the girls in Honduras. "Hel-LO, Sharon and Aggie. Goodness, am I happy to hear from you!"

"Hi Cynthia." Their two faces peered at her from the display screen.

"You two look swell. Are you enjoying your fun in the sun?"

"Absolutely! This place is beautiful. Perfect weather, white sand beach, great food, bars, and a pool," Sharon enthused.

"And a good-looking sexy CIA agent," added Agatha.

"I thought you might appreciate the bonus boy-toy," said Cynthia. "So you've had a chance to enjoy the paid vacation, huh?"

"Yes, be sure and give Mike our thanks," said Agatha.

"You want me to tell him you're on the job and working hard, right?"

"But, of course, working hard."

"Does that mean you have news?" Cynthia asked hopefully.

"Well, if no news is good, then you might say we do."

"None?"

"The Air Force did a recon—no luck. And the Space Force covered the area with a fine-toothed camera. They found nothing. No sign of a missile launch site. Not even a clearing."

"You're kidding me! How did you do that?" Cynthia was incredulous.

"What do you mean? We didn't do anything."

"I begged. I pleaded. I called in all my markers and couldn't get the CIA to budge," Cynthia whined. "They wouldn't do or say a darned thing. As far as anyone knew there was no activity within 500 miles of Honduras. You're down there twenty-four hours and you've already got action by the Air Force and the top-secret Space Force, for crying-out-loud."

"But they didn't find anything so what good did it do?"

"It told you a lot."

"Like what?"

"Well, this is just speculation," said Cynthia, "But, they've probably got some kind of portable launcher and it's well hidden."

"In the shipping container?"

"Maybe. You've got to get back and look inside that container again while the launcher is in town."

"Oh dear, that will be difficult."
"And dangerous."

Steve's Day Begins

Steve was up early. He had several errands to run before meeting with his new team of Americans. First he needed to check-out of this hotel and into a different one. It was a nuisance to have to move every week, but posing as a playboy tourist, he changed home base as a safety precaution. So far, he had stayed in more than half the hotels, bed-and-breakfasts, rental rooms and fleabags on the island, without having to repeat one. *Maybe I'll move in closer to the girls,* he thought. *Why not live it up at a five-star resort courtesy of Uncle Sam?* He chuckled to himself. This way he could keep his eye on them and it would be closer to the Blue Roof Resort, as well. Maybe he could get a room with a view. With luck he could watch Sue Lynn with binoculars.

After getting settled in his new room, Steve geared up to leave. Something told him to be prepared for a busy day. He took pains to check his weapons and make sure everything was in working order, his cell phone completely charged and his tools all in place where he could reach them.

Steve grabbed breakfast-to-go from the café and left to check in with his contacts on the island. Driving his moped he would swing by the alley behind The Pigsty, first, to see if there was any activity.

The sun was breaking through when Steve pulled up behind the Jacaranda tree and adjusted his binoculars. *Whoa, what's this?* He wasn't the only one up early. Sue Lynn and her minions were swarming like termites. This was the break Steve needed. He took a moment to make sure he and his bike were totally concealed before he settled down to watch and snap pictures.

It soon became clear that the terrorists were loading the same heavy pieces into a rental truck. This time Steve would follow, but could he do it on a moped? No telling where they would take the truck, what the driving conditions would be or how fast and far they would go. Glad that he had filled up with gas, still Steve wished he had a heavier, faster vehicle, a Jeep or a full-sized motorcycle. Steve messaged a quick note to his CIA contact in DC, asking for another flyover. Should he contact Tom and the girls? No time for that. The truck and the van were leaving.

Steve tapped the icon to follow the tracking device on the van and mounted his bike, prepared to follow but stay out-of-sight. As they moved through town on the main highway, Steve was keeping up. Although the truck and van went faster than Steve's moped on the straightaway, he caught up when they stopped for traffic lights. He worried about what might happen when they got on the open road. Driving with one hand while watching his cell phone's display screen might take a bit of "doing." Well, it was one of the bigger islands, but there couldn't be that many miles of back roads, could there?

Steve watched the street signs as they progressed, trying to memorize the route. It wasn't long before they turned off the main drag and headed toward hillier country. Steve could tell they were climbing as every so often he whizzed by an opening in the trees where he could see off into the distance. There were fewer houses around and very little traffic. Steve began to worry. He wanted to pull off and watch his tracking display but was afraid he might lose them.

Then he remembered the team button he had put on his menu. It was time to call in the cavalry. With one hand and a thumb he switched his display, taking his eyes off the road for just the wrong instant. His front tire hit a pothole in the road and sent him careening wildly. He grabbed for the handlebar. The phone went flying. The bike slid sidewise. He put out his foot to save himself, just in time. *Ouch!* Steve let off the gas and skidded to the roadside. *Oh shit! Dammit anyway!*

Back in the honeymoon bedroom, Tom sipped his coffee and glanced at the headlines. *Ah this is bliss,* he thought wondering what took him so long to get married. With one hand he reached over and turned on his phone. He set the coffee down on the bedside table and opened up his computer just to check the headlines. Having sworn off email and computer games during the vacation, this would only take him a second. Come to think of it, why did he care about the headlines? He closed the computer and firmly set it aside, determined to enjoy this day. Picking up his coffee he half-closed his eyes and listened. Outside his window a gentle breeze stirred the palm-fronds. They rattled to accompany bird songs, insect calls and the rhythm of the surf. *Ah,* he breathed, feeling the silky sheets on his legs and thinking of his wife in bed with him. *Maybe she'll come back.* The corners of his lips moved upward and his member stirred. *Later,* he thought. *Give her a rest.* He breathed in the aroma of tropical flowers mingled with coffee and took another sip.

The pinging of his phone interrupted his thoughts. Tom picked it up and gazed at the display. It was Steve's number. *What now?* he thought, thumbing on the sound. "Hello?" The response was screeching, grinding, scraping sounds, a loud crash and then silence. "Steve?" "Steve?" *Oh shit, what the hell happened?*

Sharon's voice broke in, "Hello Steve?"

"No, it's me, Tom. Did you hear that?"

"Yeah, what's happening?"

"I don't know. I just picked up the phone and heard all this noise."

"Is anyone else on the line?" Sharon waited.

"It's just us," said Tom "and whoever else made the call."

"Steve must be in trouble."

"Oh shit."

"I think his phone is still on. Let's listen. Maybe we can hear something."

Uneven footsteps sounded. Steve was limping. "Dammit, I did something to this foot," he cussed under his breath.

"Did you hear that?" asked Sharon.

"Yeah," said Tom.

"Who was that?"

"A man's voice, I think. Couldn't tell."

"Maybe it was Steve."

"Maybe he's been in an accident."

"He's lost his phone."

"Let's try shouting. Maybe someone will hear us."

"Okay, go ahead."

"Hello, hello. Is anybody out there?" Tom shouted.

Kelly came running into the room and came to an abrupt halt at the end of the bed.

"Hello, hello, anybody," shouted Tom into the phone.

"What's going on," asked Kelly with alarm.

Tom waved at her and put up a hand for silence. He listened as Sharon screamed into her phone. "Steve? Steve? Are you there? Are you there, Steve?"

"Wait!" said Tom. "Someone's coming." They heard footsteps. "Keep calling to him, Sharon."

"Hello, hello, Steve? Hello Steve," she repeated. The steps grew louder. "Over here, Steve. Steve, it's me, Sharon. Over here, Steve." The steps stopped.

"S-Sharon?" Steve's chest heaved. "It's me," he panted.

"Oh thank God, Steve, what happened?"

"Wreck," he gasped.

"You were in a wreck?"

"Yeah."

"Are you hurt?"

"Foot."

"You hurt your foot?"

"Yeah, my foot." He was beginning to catch his breath.

"We'll come and get you," said Sharon without hesitation.

"Not yet."

"What happened?"

"Chasing the damned truck. Hit a pothole."

"Okay, you were following a truck and hit a pothole. Right?"

"Yeah. Not just any truck. Their truck."

"Okay, listen, Tom is on the line. We'll get Aggie and come after you. Where are you?"

"Route 55, going up in the hills. I think we've got some time. I'll get out of sight. Look for a cairn on the right side of the road. Okay? And listen, you can't come on those flimsy motor scooters. Get a jeep or an SUV, something sturdier with more power and speed. And bring something to wrap up my foot. Okay?"

"We'll be there. Leave your phone on."

"Got it." Steve clicked off and set about to build a cairn and finding a hiding place. His foot was killing him, and so the first thing he did was find a sturdy limb to serve as a crutch. Making sure the phone was secure in his pocket, he limped back to the moped, dragged it off the road and laid it down behind a thick palmetto along the left side of the highway. This would be a good hiding place for him, as well. He began finding rocks to build a cairn across on the other side.

"Tom, listen, I can get the car and pick you up, okay?" said Sharon, still on the phone with Tom.

"Good, and I'll get the bandages and pain medicine. Also, bring all your weapons."

"Will do. See you in twenty minutes."

"Okay, leave your phone on."

"Got it."

Kelly had heard all of this. While Tom hurried into his clothes, she left the room to get bandages and a wrap for Steve's foot. She hastily tossed everything into a bag along with energy bars and bottled water.

Sharon grabbed Agatha and brought her up to date. Together they charged down to the hotel lobby and up to the concierge. "What's the fastest way to get a rental car?"

He handed her a brochure. "Take your pick."

Sharon pointed, "This one, if it's handy. I don't have time to wait."

The concierge pushed a button on his desk control-panel phone. "Bring a Jeep four-wheeler around front, right away, please." Outside a supervisor tossed a set of keys to a boy who took off on a dead run to the hotel's back storage lot. He lived to drive cars, and none too carefully either.

The concierge held out his hand, "Will that be charged to your room, ma'am?"

"Yes please." Sharon handed him her room card.

He swiped it through his machine and handed it back. "Here's your card and receipt. Just show this to the doorman out front. "Maps, GPS, and everything you need are in the car. Can I help you with anything else?"

"What is the fastest way to route 55?"

"Take a right out of the parking lot. Follow the main road five traffic lights and turn left."

"Thank you."

"You are welcome. Please drive safely and enjoy your day."

Sharon and Agatha hurried to the front doors which were swung open by two smiling young men in spiffy uniforms. "Good morning, ma'ams," they said in unison. "Your car is on the way." Their Jeep screeched to a stop directly in front of them. The boy jumped out and ran off. The two uniformed bellmen held the car doors open with a flourish.

"Thank you," said Sharon as she offered the closest one a tip.

"No, ma'am, we do not accept tips. Thank you. Just enjoy your beautiful day in paradise." He closed her door after taking care to see that she was safely belted in.

Agatha received the same courtesy. "Thank you," she said to the man and to Sharon. "Let's roll."

Five minutes later, "Climb in," said Agatha opening the door for Tom and Kelly who were already waiting outside their hotel. They barely had time to close the door when Sharon took off. "Buckle up," said Agatha. "We're going down the shore drive to the fifth traffic light and turn left on route 55."

"Have you called Steve?" asked Tom.

"Not yet."

"Okay, I'll let him know we're coming."

Steve heard his phone. "Yeah," he breathed.

"We're on our way, buddy. How are you doing?"

"I'm alive. Hidden behind a palmetto."

"How is the pain?"

"Hurts."

"Are you bleeding?"

"No."

"Good."

"Be careful driving."

"Ha-ha. You're kidding, right? I let that crazy woman drive."

"Big mistake, man. Hang on."

"You, too."

Sharon interrupted, "Heh, you people. I'll have you know my major in college was Drivers' Ed." She accelerated through a traffic light as it turned red. "Oops, that was close."

"Where did you say you went to college?" asked Tom from the backseat.

"Never mind that, hot shot. You just do your job, I'll do mine," she retorted as she swung around a taxicab, blowing her horn.

Tom reached out to take Kelly's hand. She squeezed back.

"Route 55 quick left," warned Agatha who was riding shotgun.

The Jeep took the corner on two wheels, righted itself and leaped ahead.

Tom spoke to Steve, "We just turned onto route 55. How long should this take?"

"No more than ten-fifteen minutes."

"We'll be there in five. Have you seen any traffic?"

"No, none."

"Okay. Let me know when you hear us coming."

"What are you driving?"

"A new Jeep, four-wheel drive."

"Got it. Keep the pedal to the metal."

"Don't say that, man. We're already flying."

"I'd laugh if it didn't hurt so much."

Five minutes later, "I hear you coming, man. Slow down on those hairpin turns," said Steve.

Sharon made a small adjustment.

"You've got to be getting close," Steve announced. "Shit, you missed me, man!"

Sharon applied the brakes and threw it into reverse.

Steve pushed himself up and leaned on his makeshift crutch. Tom was the first one out of the Jeep and running to help. "Hey man, let me help you."

Steve smiled through the pain and joked, "What took you so long?"

"Couldn't get that lazy driver to move it," he replied. "Which foot hurts?"

Steve pointed to his right foot.

Putting his shoulder under Steve's arm, Tom asked, "Can you move with a little help?"

Steve nodded just as Agatha ran up and threw her shoulder under the other arm. Together they moved up the side of the ditch onto the road and into the front seat of the vehicle. Steve reached down and moved the seat back as far as it would go. His leg was still sticking out.

"Hand me that bandage, please, Kelly," said Agatha.

"Here, let me help," suggested Kelly. "Can you hold the leg while I wrap it up?"

"Good idea," said Agatha. She helped Kelly as they removed Steve's shoe and gently probed the ankle. "I don't feel any broken bones," she said. "Let's hope it's just a sprain."

Kelly had torn a sheet into strips. They wrapped the foot and ankle and tied it with a narrow strip. "Okay, how does that feel, Steve?"

"Thanks," said Steve. "That should help." He eased the foot into the car. Agatha pulled the seat belt over and handed the end to Sharon who snapped it in place. Agatha carefully closed the door and piled in the back with Tom and Kelly.

"Did you get my moped?" asked Steve.

"It's in the back," said Tom.

"Thanks, you guys," said Steve. "Let's drive on up this route and I'll fill you in as we go."

Sharon put the car into drive gear and started forward a good deal more smoothly than she had been driving on the way up. Steve was able to watch his cell phone with both eyes now. The tracking device was working perfectly. It showed that the van had stopped. Presuming that the truck and van were together, the USA team pulled the Jeep off the road into a brushy hiding spot about 500 yards out. Agatha and Kelly stayed back with Steve.

Tom and Sharon took the binoculars and set out on foot. Before separating they had a considerable debate about who would go and who would have the tracking phone. Steve won that argument when he pointed out that the phone only recognized his face and thumbprint. Besides he needed to know if the truck and van were coming, whereas Tom and Sharon could hear and see them in time to take cover.

Once underway, the recon team of Tom and Sharon, had no trouble identifying the target by the banging, crashing, shouting and cursing echoing through the trees. Nearing the activity, they crept behind cover and took turns watching with the binoculars.

"What do you see?" asked Tom.

119

"I think they are putting up a missile launcher, don't you?"

"Yes," said Tom in low tones.

"Can you get good pictures from here?"

"I wish I had a better camera."

"Well, let's separate and move in closer. We'll try to get pictures from more than one angle."

"Good idea. But, we need to keep radio silence, so let's set a time limit and meet back here."

"Okay, see you in twenty minutes."

Sharon crouched low and moved into the woods on the left. Tom went on the right. Twenty minutes later she returned to the spot and hid behind the shrub. She looked around to ascertain it was the right place. Glancing at her watch, she settled down to wait, feeling confident the pictures she had obtained would tell a lot.

Five minutes passed before she checked her watch again. Where was Tom?

Three minutes later Sharon was becoming concerned. What if Tom was lost? What if he got caught? She looked around for another hiding spot, just in case he showed up with a guard holding a gun on him. Quickly Sharon crouched and moved to a bush farther away. She could still see the first bush, in case he came. She decided to give him five more minutes, max, and then she would call Agatha to make a rescue plan. They couldn't leave Tom behind.

Minutes ticked by, then seconds. Sharon lifted up her phone to call when she saw movement. Someone was sneaking through the brush. *Oh my, it's Tom.* Was he alone? She watched him creep close by her hiding place. "Psst," said Sharon. He did not look up. She picked up a stick and threw it across his path. His head jerked. "Psst, over here Tom." He looked her way but did not see her. "Behind this bush. Turn left, twenty feet."

Tom slid in beside her and sank down. "Wow am I glad to see you."

"Likewise."

"I got pictures," she said, holding up her phone.

"Yeah, me too."

"Let's go."

They wasted no time skulking through the bushes as fast as they dared until they thought they were far enough away to come out in the open. Even so, they were careful to stay off the road, in case the terrorists had posted a sentry. It took some time to work their way back to the Jeep.

The reunion was met with huge relief on all sides.

"Thank God you're back," said Kelly.

"Were you worried?" asked Tom.

"Is the Pope Catholic?" asked Kelly.

Tom climbed in next to her and gave her a big hug and kiss.

"No need to worry, sweetheart," said Tom reassuringly. "We were never in any danger."

"What did you find out?" asked Agatha, wishing they could get going.

"It's a portable launching platform, all right," said Sharon.

"I got some good pictures," said Tom scrolling through the pictures on his phone. "Here, have a look," he said handing the phone to Steve in the front seat.

Steve swiped the display and whistled. "State of the art. I've never seen one quite like this."

Meanwhile, Sharon was looking at her display. "Here it is from another angle, Steve."

Steve handed Tom's phone back to him. "Send those over to my phone, please, Tom." He grasped Sharon's phone and scrolled through her pictures. "Same thing, different angle. Good idea," Steve said.

"Shall I send those over to you?" asked Sharon.

"Never mind. I can do that. Why don't you drive?"

"Oh sure," said Sharon, starting the Jeep and backing out of their hiding place, while Steve worked on the phone.

"Let's drive on down to The Pigsty," Steve suggested. "Maybe we have time to get into their computer lab and poke around before they come back."

Kelly could feel her stomach knotting up, but she bit her lip.

"Okay, sounds like a plan," Sharon enthused pulling onto the road and heading downhill.

"While you drive, I'm going to put these pictures into one file and send it up to headquarters. They should have some overhead shots from the airplane soon, too."

Chapter 9

The Pigsty Gift Shop

Sharon circled three times to make sure no one was around. Satisfied no cars were parked in the alley behind the computer lab, she pulled the Jeep into an open space on the street in front of the store. Agatha would reconnoiter the front door to see if the store was open. If Agatha went inside Sharon would drive away planning to cruise by every sixty seconds until Agatha came out of the store.

This proved to be unnecessary. The place was locked up, a shade was pulled and a Closed sign was hung in the window. Agatha returned to the Jeep and reported, "The store looks deserted. So what does anybody want to do next?"

"Let's hit this place," declared Steve.

"We'll never get a better chance," Tom agreed.

"Okay," cautioned Kelly, "but we need to be careful."

"Everyone: make sure your phones are on," advised Steve. "Have you all got a 'quick call' icon on your display?"

"I don't think so," Kelly questioned. "What is it?"

"Hand me your phone, and I'll fix it for you," Steve offered. "Let me see if you have all of our numbers entered in your contact list." He quickly checked. "Okay, you do. So now we want to put all four of us under one icon so that you have only one button to push to call everybody at once. It works like an intercom or communicator." He tapped a few more times. "Here you go, Kelly. All fixed. Now see this smiley icon on the display? That's your 'quick call' button for all of us. The minute you're in trouble, just make sure your phone is on and hit that button twice. That's what I did when I got in that wreck. Tom and Sharon heard me and you-all came to my rescue, thank goodness."

"One of us needs to stay with the car," stated Tom.

"Well, clearly Sharon is the best driver." Everyone laughed.

"What's so darned funny," Sharon challenged. "Are you impugning my driving skills?"

"Your skills are right up there with the Andrettis'," assured Steve, coming to her defense and suppressing a grin.

"I think any one of us can drive," Agatha observed, "except for Steve who has his foot wrapped up. The more important consideration is who can pick a lock, and who can hack into a computer?"

"And we need two lookouts, one in back and one in front," said Sharon. "That leaves two people to go inside."

"Who is the best at computers?" Steve wondered.

"Well, Kelly's degree is in computer science, if that helps," boasted Tom.

"I nominate Steve for his lock-picking prowess," cracked Sharon.

"And I think Sharon and Agatha are the best bodyguards," quipped Steve, recalling his experience. "I think I can manage with my handy-dandy crutch." He pointed at his stick.

"If you're sure, I guess that leaves me to drive the getaway car," observed Tom.

"Let's make sure our phones are on vibrate. I'll call you now so everyone's line is working." He tapped the team icon twice. "Report in please, everyone."

"Sharon here."

"Aggie here."

"Tom here."

"Kelly are you online? Answer please," said Steve.

"Oh, I'm sorry," said Kelly. "I guess I'm on."

Steve frowned and checked his watch. "Okay, Kelly, let's go. Time's a-wastin'."

"Oh, I'm not sure about this," she demurred.

"Go ahead, babe, you can do this," Tom cheered.

Kelly nodded and slowly opened her door. Sharon was already out on guard duty. Steve was hobbling along and halfway there. "Go on Kelly, you need to get started," urged Agatha. To Tom she asked, "Do you want to drive me around back, or shall I go with them?"

"Maybe the car should stay here where I have a good view. You stroll around back. Here, take these binoculars and hide behind the Jacaranda tree. Stay concealed. Then if you see anything, send up the alarm."

"Okay. Have your phone on."

"Got it. It's on and the line is open."

Steve had already picked the lock on the front door and disappeared inside. Sharon was ambling the sidewalk, phone in one hand, pretending she didn't see Steve, watching the passing cars, alert and ready for anything. Agatha left and headed in the other direction.

"Okay, Kelly, honey, go on, the front door is open. Go inside and lock the door behind you," Tom directed.

Kelly left the car, head down, giving herself a pep talk. *You go girl. Stop being a wimp. Remember who you are. Kelly the curious, right!* She lifted her chin up and strode to the gift shop door. *Pretend you own the place.* Hiding her fear, she opened the door, slid through and locked it behind her. Steve had vanished and so she made her way to the door marked "Employees Only- No Admittance," in English and Spanish.

125

No overhead lights were on inside the computer lab. Kelly waited a few seconds for her eyes to adjust. *I'll remember this if someone comes. They won't be able to see, at first. Will they?* Kelly approached the first computer and jiggled a nearby mouse. This one seemed to have an old-fashioned keyboard, except there were no markings on the keys. She jiggled the mouse again and waited. Nothing happened. Feeling for a starter button, she tried that. Still nothing. Wasting no time, she moved down to another computer that seemed to have no controls. She tapped the display. A picture opened up. Kelly tapped in the center. A small window opened with a place to enter an ID and password. Kelly left that and moved to an iPad. She tapped the display. Another similar window asked for the ID and password. She tapped "forgot ID". A voice told her in Spanish that a code would be sent to her phone. A window opened where she could enter the code.

Kelly felt someone brush her arm. Adrenaline shot through her body. She jumped and screamed.

"Quiet!" said Steve. "It's only me."

"Oh my God, you scared me to death."

"Calm down, Kelly and help me figure this out."

"Figure what out?"

"This cell phone I found on the desk."

"Well, you're the cell-phone genius. Why don't you tap 'Messages'?" she mocked, still a bit miffed over the scare he gave her.

"Heh, I said I'm sorry."

"You did? How did I miss that?"

"I'm sorry, I'm sorry." Was he really sorry he scared her, or just sorry she was here?

"Try harder."

"I'm sorry, I really am sorry, Kelly, please forgive me," he pleaded, overstating the case.

126

"All right just don't do that again," she snapped.

"Got it." Steve vowed he would not touch her again if his life depended on it. "This looks like the code you need," he said, gazing at the phone.

"Can you read it off, please?" said Kelly politely.

"RQN7000401T."

"RQN … can you read the rest slowly, please?" Kelly was tapping the display keyboard.

"RQN 7 0 0 0 …"

"Got it. RQN seven thousand. This display keyboard is funny."

"What's wrong with it?"

"The keys are all blank."

"Really? I wonder why."

"Well, it's okay. I remember them, I think. So what's the rest of the code?"

"4 0"

"4 0" repeated Kelly.

"1 T as in Thomas. That's it. Why don't you read it back to me?"

"Okay, here's the whole thing, RQN7000401T."

"That's right," said Steve.

"Now I need the password," said Kelly. She had entered the ID and clicked 'forgot password' on the screen. "Watch for another new message, please, Steve."

Steve waited.

Finally the phone peeped again. Steve tapped 'Messages'. "Ah, here it is." Steve opened the message. "Okay here we go. Um … do you read Spanish?"

"A little. Let me see."

Steve held the phone for her.

Kelly squinted at the phone. "It's so dark in here."

"Yeah."

"It looks like the code is in symbols, not letters and numbers. May I take the phone, please."

Steve handed her the phone.

Kelly held the phone with one hand and tapped with the other. "Okay here we go. It's allowing me to change the password. Shall I do it?"

"Boy oh boy, I don't know. If we change the password, they're going to know we were here, or someone was here, aren't they?"

"Not necessarily. They may just go through the same thing we did and change the password again, especially if you can delete those messages. They probably won't know what happened."

"You're right. Okay, change it to 'teamUSA5'."

"Okay."

Kelly typed in the new password once and then typed it again to confirm. She hit enter. "We're in!" Kelly was jubilant.

"Now what?" asked Steve.

"Let me open some of the documents," proposed Kelly, "while you read some of their phone messages. Maybe we can figure this out."

They both worked away on the devices.

At length Kelly reported, "Steve, everything I see is in Korean!"

"Oh no! Are you sure?"

"Yes, I recognize those symbols from the side of the shipping container, and also on the missile."

"That explains why I can't make sense of any of this either."

They gazed at each other.

"Any ideas?" queried Steve.

"Maybe a couple."

"Good."

"We should copy this onto a flash drive and send it up to headquarters. But, also I can ask the computer to translate it for me."

"I'll look in the desk for a flash drive," said Steve turning away and hobbling over to the desk.

Kelly started looking for the translate command.

Just then both their phones vibrated.

"Oh-oh," said Steve, reaching in his pocket for his phone. "There's a call coming in, Kelly. Have you got it, too?"

She nodded and reached for her phone. She tapped the phone icon and put it up to her ear. Agatha's voice spoke, "The truck and van are on their way. You've got twenty seconds to bail out the front door."

"Oh my God!" cried Kelly, terror gripping her. She turned to flee.

"Wait," ordered Steve. "Here's a flash drive."

"No time!" Kelly shrieked.

"Do it!" He thrust the drive into a USB port. "Tell it to copy."

Kelly's hands were trembling so badly she couldn't type.

"Type!" commanded Steve.

Kelly grasped her right wrist in an attempt to hold it still and tapped with one finger. 'File' … 'Save-a-copy' … 'Browse' … 'Removable disk D'. The commands were written in Spanish, but she figured it out. She watched wide-eyed as the line crept agonizingly left to right and stopped. She hit the back arrow and then 'File-close' and 'Power options/Sign out'. Her teeth were chattering, her face white as a sheet.

Steve coolly pulled the flash drive out of the port and secured it in his pocket.

"They're here, guys," Agatha warned.

"Your Jeep's at the front door," stated Tom. "Hurry!"

"I've got you covered," assured Sharon.

"Run!" screamed Agatha.

Without a word Steve grabbed Kelly's arm and pulled her away. "C'mon, Kelly, we're okay," he said calmly while all-but carrying her on a lop-sided run. Reaching the front door, Steve shoved her through, locking the door behind him. Tom held the Jeep door open. Later Kelly would not remember how she got in. Doors slammed and the Jeep moved into traffic.

"I'll pick you up, Sharon," Tom spoke into the phone.

"No, I'll keep watch a little longer," she replied. "I'm safe here."

"Same here," said Agatha.

Kelly collapsed.

"You did it, girl," said Steve. "Congratulations."

Operation JHG

Deep inside a mountain on the other side of the world, several crack teams of nuclear scientists, engineers and physicists were being supervised by military commanders under the supreme commander General Lee.

Lee had been applying every discipline he knew to force the workers into more and faster production. Whenever he met with DL—Dear Leader—he was a mousy subordinate. But, inside the mountain he was a different man, autocratic, ruthless and uncaring. Of course, DL had chosen him to head-up this mission for those precise characteristics. DL needed someone who could get the job done by whatever means necessary.

The scientists and support personnel were totally cowed under a twenty-four-hour guard. They dared not whisper or even think of insurrection. They had seen whole families disappear. People could be whipped for the smallest offense, taking too long in the shower or walking too slowly. Being sick was not permitted. No one was safe. Once, even the chief officer, the head of operations, vanished without a trace. One day he was there, strong and completely in charge. The next day he was replaced with another man as if nothing had happened.

They were not told the truth about what they were doing. The workers had their private ideas but no one dared question or say a word. One team had designed a small light-weight satellite for which purpose they were given a ridiculous explanation, being told it would broadcast children's programs. Three prototypes had been handmade, but the workers had no idea where they had gone. The designers were rewarded with a medal ceremony and a trip to the train station where they thought they were headed home. Instead they were taken to forced labor camps. The mission was too important to risk allowing these people out into the general population where they could talk. Perhaps someday they would be released, if they survived and if they were not forgotten.

Now, the pressure was on to design small atomic explosions that would weigh just a few pounds each. Supposedly these designs would be manufactured and sold all over the world to mining companies and bridge construction outfits for the purpose of excavating large earthworks. It was important that the devices be controlled remotely from large distances for safety reasons.

One way that General Lee maintained secrecy was to separate the mission into individual parts. Certain groups worked on one aspect of the design without interacting with the others. Each group lived apart, worked and took their meals together, completely separated and barely knowing the other groups existed. A different group had designed the portable rocket launcher from the group who had designed the rocket, itself. A third group had designed the satellite.

Now, a fourth secret group was working on a device to attach the bomb to a satellite. This group had the third satellite in their detached unit, separated from the others. They were given an empty container to attach to the satellite, never knowing that the container would one day hold an atomic weapon.

Perhaps the most highly skilled secret group was working on the weapon. Without the weapon all else was useless. Unfortunately, this group was not making the required progress. General Lee was on the horns of a dilemma. He could go only so far in pushing these people without destroying them.

Accustomed to ruling by fear and with an iron hand, his motivational people-skills were almost non-existent. And so his frustration was mounting exponentially. He compensated by taking it out on the little people, those servants who did the cooking, cleaning, washing and pressing. Hardly a day went by without a public flogging taking place during mealtimes or outside the dormitories in the middle of the night, when everyone could hear the screams.

Conspiracy

Throughout history there have been tyrants who have risen to power and managed to survive through bribery, greed, fear and ruthless intimidation. Correspondingly, there have been hordes of sheep who follow, content to merely survive and lick up the few crumbs that come their way.

Fortunately for human survival, there have been tiny points of light, those special individuals who have not been completely cowed, who have continued to have a spark of individual thought. Such a person worked for General Lee on the small atomic explosive project. Kim Jung Lo had never forgotten where he came from. He dreamed of the day when he would return to his family and the farm where he grew up. A bright young man, he had done well in his studies and been singled out for advanced training. The farm work had turned him into a strong, healthy, well-fed youth. But the long months working inside the mountain, living on meager rations had weakened his muscles and turned his body into a pale, thin replica of his former self.

There had to be a way out. But to Kim Jung Lo, finding a way to sabotage this project became more important than his own safety. Jung Lo was no fool. He could see what was happening here. Much as he longed to go home, he had to take a chance. His country's future and the world's survival hung in the balance. Jung Lo kept his thoughts to himself and continued pretending to work on the design of a small atomic explosive as he worked his way into the position of supervisor of the design and testing phases of the project. As such, he watched for small ways he could manage to delay progress. It was only a few seconds here, a minute there, but it added up.

At the same time he gave General Lee glowing evaluations of their progress. At last he was able to report, "We will be ready for a test-firing next week, sir."

"There will be no test-firing," Lee responded.

No firing? Jung Lo was stunned, but wisely held his tongue. Instead he waited for General Lee's explanation.

"Dear Leader has ordered us to prepare two devices immediately. Your orders are to make sure it will fire by doing computer simulations. There is no need for a test-fire. That only wastes time and materials. The device will be tested when it is put to use in a mining operation."

This was the break Kim Jung Lo needed. He loved computers and spent every spare minute working on various concepts and designs. They were already so close, it would be a simple matter to design a small atomic weapon, but this one would have a fatal flaw. The difficulty would come in fooling his fellow designers. From that point on, Jung Lo spent hours going over prototypes of his own design and those of his team, testing them in computer simulations. Lying on his bed at night, he went through them in his mind, devising various ways he could cause the device to misfire. It must be undetectable until the actual command was given. As with the main device, this Trojan horse would not be tested. It would have only one chance to work, or to fail.

Jung Lo finally decided that he had to insert an error into the thousands of lines of code that would control and detonate the bomb. This error had to look innocent if ever discovered. Someone could see it and might attempt to fix it but Jung Lo was sure no one would dare do so without consulting him. If that happened he had a backup plan.

Washington D.C.

CIA headquarters was *not* amazed at the pictures that came in from Honduras. Even less telling were the overhead shots taken from the spy plane. There were much larger disasters, emergencies and crises going on all over the world in Afghanistan, Syria, Russia, Turkey,

China, Korea, Europe, North Africa, Venezuela and many other places. These pictures were just another batch of information from the mountain of data waiting to be reviewed.

Mitch Mucurie was a fairly new young CIA analyst assigned to the more routine tasks. His job today was to receive and examine pictures as they came in from the spy-planes, and sometimes from other sources.

When the spy-plane's pictures arrived, Mitch took measurements and made a schematic drawing of the Honduras launch site and the various vehicles and personnel. He added facts that the plane supplied such as GPS position, time, date, weather and terrain. A map showing the location went into a file with the other data. Later in the day when he received the pictures from Special Agent Steve Spalding, Mitch was able to add a great deal more data to the file about the people shown in the pictures. He gave each one an identifying number, marked the pictures accordingly, and made estimates of their height, weight, sex, and age. All this was combined into one file awaiting Agent Spalding's written report. For the time-being the file joined a large stack of other reports in Murcurie's pending file. When complete it would be assigned to another more experienced analyst's desk for review and evaluation. CIA dealt with hundreds of similar cases every day, representing thousands of bits of data.

The Island Hotel

Meanwhile, on the island, Special Agent Steve Spalding and the USA team debated what to do with the cell phone and the flash drive that he and Kelly had retrieved from the computer lab.

The five of them were gathered in Steve's room after their exciting day. Steve parked himself in front of the computer, his ankle propped up on a chair, wrapped with two bags of frozen peas and covered with a towel.

"Well," suggested Tom, "I think we need to put this drive into a computer and see what we can make of it."

"Sounds good to me," said Sharon. Agatha nodded agreement.

"Kelly?" Tom inquired, turning to look at her.

"Oh … oh sure, go ahead." Kelly had collapsed into the softest easy chair in Steve's hotel room, clearly uninterested in what was happening. She sagged back and closed her eyes.

Steve noticed her condition for the first time. "Sharon, would you mind looking into my kitchen for something for Kelly, maybe some tea or an energy drink?" Sharon took one look at Kelly and left immediately. "Tom would you please get that throw off my bed and cover her?" Steve suggested.

Tom grabbed the throw and tucked it carefully around his wife. "You were fantastic, darling," said Tom. "How are you feeling, sweetheart?"

A feeble 'Thanks' was all she managed.

Sharon returned with a chocolate bar and energy drink. "Here you go, Kelly. Have a little of this." Turning to Steve she asked, "Where do you keep your pain pills?"

"I'm okay, Sharon."

"Yeah, I know, but where do you keep your pain pills?"

Steve nodded down the hall. "Check the bathroom."

Seeing they had things under control Steve inserted the flash drive into a USB port. His computer recognized the drive and pulled up a list of files, all in Korean. Steve chose one file and asked his computer to open it. "Hey look, guys, my computer has software that will open this thing."

Sharon returned with two anti-inflammatory pain pills and a glass of water. "Here," she held out her closed hand.

Without taking his eyes off the computer, Steve opened his hand. Sharon dropped two pills in his palm. Steve tossed them in his mouth. Shoving the water into his hand, she said "Drink." He took a gulp and held the water glass out to Sharon. She set it aside and rested her chin on his shoulder gazing at the computer screen.

Tom and Agatha moved up behind Steve and watched over his shoulder. "Ask it to translate the document, Steve," suggested Agatha.

"Not sure it can do that. I don't have every language." Steve had a powerful computer, with a mouse, a regular keyboard and a full array of programs. He searched through his languages and shook his head. "I'll have to go online," he said. "Grab that bag, please, Sharon." He pointed toward a brown leather zipper bag. "Reach in there for a couple of things, please. First, hand me that Ellipsis device and then see if you can find a charger that will work on this telephone." He reached in his pocket for the cell phone he had brought from the terrorists' lab.

Sharon handed him the Ellipsis and continued digging through the various cords, connectors and devices in the bag. She tried a couple of different chargers, found one that worked and plugged in the cell phone.

Steve took the small device and began booting it up. "This little baby will get me a secure connection," he said. "I wouldn't dare use the hotel's Wi-Fi." The Ellipsis came to life and displayed a four-letter code. Steve set the device on the counter and turned to his computer. He asked the computer to show a list of available networks. It showed a long list of hotel networks and other private ones nearby. Steve clicked the Ellipsis choice and entered the code. In seconds he was on a secure line. The address line was ready. Steve typed 'translate Korean to English' and several choices appeared. Steve selected one and a window opened inviting him to enter the Korean words. Steve moved his cursor over and clicked back into the letter file. He highlighted it all and copied it. Moving back over to the translator, he pasted the file into the Korean box. The English translation appeared on the other side.

"Ain't technology grand?" asked Steve.

"It's like magic!" Sharon enthused, and bent to read the document.

The four of them crowded around reading the message in English while Kelly drifted off to sleep.

"This doesn't look like much," said Steve. "Let me try another one."

After translating five documents, Steve shook his head and sighed. "Looks like this is going to take some time. Look at this long list of documents."

"Can you translate just the list of titles, Steve?" asked Sharon.

"Maybe we can narrow this down," Agatha added.

"Good idea," said Steve. In a few seconds he copied and pasted the first three hundred titles. "Ah, here we go," he said as he scrolled down through the list. "What stands out to you?"

Sharon pointed, "Try this one."

Steve switched windows and opened the one Sharon selected. "Bingo!" He enlarged the document and began reading, "Listen to this, gang. Target address Kansas City. Coordinates are 39.1141° N, 94.6275° W. Practice programming."

"That's all it says," observed Sharon.

"What does this mean?" Agatha wondered.

"Well, I suppose it gives the coordinates of Kansas City, USA. Is there any other Kansas City?" Sharon responded.

"Right in the center of the United States!" Agatha exclaimed.

"Yup."

"Isn't Omaha the center?"

"Could be."

"Doesn't matter that much, does it?"

"Well Kansas City is pretty much the center of North America."

"Right."

"The operative word here is 'target' isn't it?"

"Oh my God!"

"These people are trying to shoot a rocket at Kansas City."

"Holy Hanna!"

"But, how can they do that? I mean, don't we have a missile defense?"

Tom remembered the words of Dr. Peter Kinney on the Haven-Harbinger TV show. Tom glanced at Kelly to make sure she was asleep. He did not want her to hear this. "Well," Tom began, "actually we do and we don't."

"What's that supposed to mean?" Agatha questioned.

"It means we have a missile defense which protects the United States from a missile attack from the north, but not so well from the south."

"No way. You've got to be kidding!"

"He's not kidding," said Steve. "Actually the defense rockets can aim in any direction, but if a missile was launched from a ship or submarine, close-by, or from a border country, it would be well over the US before our defense rocket would meet it."

Agatha's mouth dropped open. "I can't believe that. Why?"

"It has to do with the political situation, I suppose," Tom explained. "There's always opposition to the military, you know. Considering that they only have so much money in the budget, the Pentagon has to place the defenses where they will do the most good. They protect us from the north, where the most likely threat is. I guess there was never enough money or will to defend the southern border."

"Russia, right?"

"Well it was the USSR, originally and now it's Russia which has many warheads and ICBMs capable of reaching the US. They could fire these over the North Pole and approach from the north."

"But, all we saw in the shipping container was a missile. Where's the warhead?"

Chapter 10

Where's the Warhead?

General Lee was demanding two devices be ready today. Chief scientist, Kim Jung Lo could delay no longer. He was up early working on the final commands to be inserted into the control computers. This would allow remote controllers to set off the explosion. It could be ignited immediately or be placed on a timer to explode in a certain number of hours, minutes and seconds.

Jung Lo ran the string through his simulator one more time. It worked perfectly just as it had hundreds of times. He scrolled through the commands and inserted two stops carefully designed to appear as simple typos. Next, he saved the file and left to join his team at their breakfast table. Moments later everyone appeared blinking sleep from their eyes. Every chair was filled except for that of Hua Ming Du, Jung Lo's assistant and second in command. It took very little time to eat through the meager rations of black bread and water.

Jung Lo was about to break the uneaten bread into equal pieces when his assistant raced up and grabbed it.

"No, wait!" Ming Du exclaimed, breathless from his haste. "I need that." He wolfed it down. "Come with me, Mr. Kim, sir."

"What?" said Jung Lo.

"I've found something you need to see," he said excitedly. "Come," he insisted.

Feigning indifference, Jung Lo rose from the bench and followed.

Together they entered the password to unlock the door to their design center. Hiding his concern, Jung Lo demanded, "What is this about?"

"I've found something you need to see?"

"What were you doing in here alone?" accused Jung Lo. "That is unauthorized."

Ming Du hesitated. "Oh...I..." he stammered. If he confessed to breaking the rules, he could be in terrible trouble. "P-please, I beg you. Don't tell the commander. I have a wife and children. But you must see this!" Tears threatened to spill from his eyes. He clasped his hands in a prayer-like position and bowed several times.

Kim Jung Lo glared at him. "And why should I cover-up for you? I have an unblemished record," he boasted, firmly crossing his arms and rising to his full height. Jung Lo knew what was coming, but he thought it best to frighten Ming Du and keep him on the defensive. This fear could assist Jung Lo in the subterfuge he was about to employ. "Stop sniveling and stand up like a man. Tell me what is so important that you have dragged me away from the breakfast table," he ordered harshly.

Hua Ming Du could barely straighten up. "Yes, sir," he said, bowing again. "I have found a flaw in our program, sir," he managed, between bows. "Please grant me permission, sir."

"All right but make it quick. I'm very annoyed that you were snooping in the program unaccompanied. How do I know you haven't sabotaged the program, yourself?"

"Oh no, no, no. I would never do that." Ming Du realized he had made a terrible mistake running to his supervisor. He should have fixed the typo and told no one.

Jung Lo opened the program on the main computer. "All right, Mr. Hua Ming Du, show me this error and I will decide whether you shall live or die for this gross infraction of the rules."

Ming Du scrolled through the program until he located the line of code that Jung Lo had installed earlier. "Here it is, sir. See this. Right here," Ming Du pointed a shaking finger.

Jung Lo pushed him aside. "Let me look at this." He spent a minute scrolling up and down, scrutinizing the 'error' and mumbling to himself, keeping the poor man quivering in terror. At last Jung Lo looked up. "I believe this is a typo. Let us assume that is all it is and not a deliberate and treasonous attempt at sabotage." He scowled at Ming Du as if the man was guilty. "Is it your opinion that this line should be erased?"

"Yes, sir," said Ming Du.

"Well, in that case you will observe as I delete the error, so that there is no question later." Jung Lo carefully deleted the entry. "Now, please look at this and confirm that it is correct."

Ming Du leaned closely, "Yes, it is correct."

"Very well, then, I shall save this and close and lock the program to prevent any further intrusions. Please observe as I do this." Hung Lo worked slowly and deliberately so that it could be seen by Ming Du. "Are we agreed now?" asked Jung Lo.

"Yes."

"You are certain?"

"Yes."

"All right, then. Let us speak of this to no one. You are dismissed."

Ming Du was enormously relieved to be let go without charges. They left the room together.

Today the program would be added to the tiny but powerful computers that would control the two bombs.

The devices would then be carefully swaddled in packing materials, loaded into two separate containers and readied for shipment to the two launch sites in Central America. There they could be attached to the small satellites, ready for mounting on the missiles, awaiting orders to launch.

The controllers would be programmed, as well, and given to General Lee.

Hotel Room

Kelly opened one eye. She could see Team USA gathered around the computer and talking excitedly. She felt hot. It was stifling in here. She moved to sit up and then realized she was covered with a blanket. No wonder it was hot. She tossed the blanket off, sat up and stretched. *I must have fallen asleep*, she thought, and then the whole day swept into her brain.

"Good morning, Kelly," said Tom, noticing for the first time that his wife was awake. "How are you feeling, sweetheart?"

"I'm fine. Sorry I crashed out on you."

"Heh, you needed it. No problem. Can I get you something, sweetheart?"

"Maybe a little juice or something. I'm thirsty."

"You didn't finish your energy drink," he said, picking it up from the table. "Let me freshen this up."

Tom moved into the kitchen area while Kelly turned her attention to the others who seemed to be earnestly looking at something on the computer display.

Sharon looked up. "Oh, hi Kelly. You're awake. Good. Wait 'til you see some of this stuff we're finding on that flash drive you got. You did a heck of a job Kelly. This is incredible."

"I did? It is? Really?"

"Yeah really," said Steve, looking up. "You hit the jackpot."

Tom handed her the iced drink. "Here you go, sweetie. Drink this. You'll feel better."

Kelly's mouth did taste like a herd of buffalo had passed through. She drank thirstily, and handed the glass to Tom, who set it on the table. She stood up. "May I use your bathroom, Steve?"

"Oh sure, help yourself," Steve pointed down the hall.

Homeland Headquarters

DL was having his sumptuous breakfast of galbi, kongnaumul bab, oi naenaguk, kimchi and moe saingchae. Each dish was served on elegant re-embroidered lace by two beautiful young women clad in flowing silk robes, using priceless fine china, crystal stemware and gold tableware. At last, DL leaned back and patted his stomach. With a wave of his hand he ordered the dishes cleared away and his coffee brought in while he visited the adjoining bathroom.

General Lee had waited an hour for DL to finish. He did not dare intrude on Dear Leader's morning routine. Sometimes DL would invite Lee to join him for coffee. Other times DL spent the time in meditation, preferring to be left alone. When ready he would summon Lee for the morning briefing. In truth, these briefings were boring, but necessary. DL had to keep the general on his toes. At last DL spoke, "Come in General." General Lee entered carrying his usual armload. But, this morning, instead of stacked files, he carried an elaborately tooled and engraved box, containing the two controllers. He planned to present those to DL with appropriate pomp. He saluted and then bowed, remaining bowed until DL spoke.

"Good morning, General" said DL without looking up. "Please place your files over there and be seated."

Lee put his box on a side table and took a seat, remaining erect in his chair.

"Please relax and tell me how my forces are doing." He yawned. "What discipline have you imposed?"

While General Lee gave a rundown on the number of floggings and the prisoners who had been shot for trying to escape, Dear Leader put his easy chair all the way back, turning on the massage function.

"Very good, General." He closed his eyes with a contented smile. "Now tell me when my little surprise will be ready for that imperialist scoundrel, Bigelow."

"I hope you will be pleased to learn that operation Jagjeon Haeg Gyeoul is nearing completion."

"How so?" DL suddenly stopped the vibrator and sat upright.

General Lee allowed himself a tiny smile. "This very day we are preparing the bomb for shipping."

"Bomb, you say? I thought there were two bombs."

"Oh yes, of course. Yes, there are two. I apologize for misspeaking."

"Go on, go on," urged DL, impatiently.

"Yes, of course, sir. The bombs are ready."

"You tested the controls?"

"They were tested on the computer simulator, as you ordered, sir."

DL cleared his throat. "Well, then, you will deliver the controllers to my desk, as soon as possible."

"Excuse me, sir, may I comment on that?" Lee asked, very carefully. He knew that DL did not tolerate any questions.

DL scowled at the General. "You dare to comment?"

"If you please, sir."

DL merely waved.

"Sir, may I say, we would be pleased to bring the controllers to you and demonstrate how they work?"

DL considered that for a moment. Perhaps there was a bit of wisdom in learning something. "General Lee, please bring the controllers here and show me how they work."

Lee considered how much he dared to say. How could he word this? "Dear Leader, I shall be more than pleased to show you this fine work that your scientists have made. The range of the controller will exceed any known explosive device."

"How is that?" demanded DL, suddenly thinking he has been the butt of a hoax. DL was always alert for any threat to his authority. A dictator can trust no-one.

"The range of the controller, sir, exceeds any known previous explosive."

"Are you saying it has a limit?"

"Any controller has a limit beyond which the beam cannot be successfully activated, sir."

"You will correct that, General," said DL, assuming that all he had to do was order something and it would have already happened. That subject was closed. Turning to a final issue, DL ordered, "You will give the scientists who have worked on this their usual reward and see that they are silenced."

"Yes, sir."

DL waved General Lee away, leaned his lounge chair back, turned on the massage function and closed his eyes.

147

General Lee picked up his box, tucked it under his arm, and left the room, with a wry smile on his face, thinking, "This is my insurance policy."

Smart Cell Phone

The USA team members were so engrossed in translating Korean documents and letters that, at first, they did not notice a beeping sound.

"What's that sound?" asked Kelly.

Everyone heard it and grabbed for their phones. They checked the screens shaking their heads.

Sharon ran to the kitchen where she was charging the terrorists' cell phone. "Oh my God, it's this one!" She gaped at it. "What shall I do?"

"Answer it," yelled Steve.

Sharon tapped the phone icon and held the phone to her ear. "Yah?" she said, waited a second and then shook her head. "They hung up," she said.

"Try the Message icon," Steve directed.

Sharon tapped a couple of times, stared at the display and said, "It's in Korean, darn it."

"Bring it here," said Steve.

Steve set the phone in front of him and turned his computer to an English to Korean translation page. Quickly he typed through the letter keys on his keyboard in order from Q to M. The machine displayed the Korean alphabet. Steve hit Ctrl plus Print and his printer came to life. Sharon handed him the printout. Steve propped that in front, next to the cell phone. Then he returned to the Korean to English page and began to enter letters into the Korean side. "I hope this works," he said. After a few clicks, he shook his head. "Nope, it just goes in as English. I'm going to have to reset my keyboard to the Korean language."

"Can you do that?" Sharon was amazed.

"Can spies spy?" he retorted, with an intimate smile, while hastily typing commands on his keyboard. A list of possible languages came up. Steve scrolled through those with his mouse and selected Korean. Then, just to make sure, he opened a new document and typed through his letter keys in order from Q to M. This time Korean characters appeared. He started a new row and typed Q to P, enter, A to L, enter and Z to M. Now he had a reasonable facsimile of the way his Korean keyboard was set up. After he ordered print, Sharon retrieved the printout, brushing his arm. Steve looked up and took the paper from her fingers, "Thank you, dear." He propped it up next to the cell phone and went to work. It took a while to look at each character, one at a time, and find the correct key to strike. "Sharon, would you please look at this and compare it to the cell phone message. Have I got it right?"

Sharon took a minute to compare. "Looks good, Steve."

Steve copied the paragraph onto his clipboard, moved into the Korean to English translation window and pasted it into the Korean side. This time the letters appeared in Korean.

In a jiffy the English translation emerged. "Joilè! Steve, you are amazing," said Sharon. "Look at this, guys!"

"Oh my goodness!" said Agatha.

Kelly read it aloud, "Device is ready. Shipment going out today. Watch for arrival."

"What does this mean?" asked Tom.

"Well," said Steve, "I'm guessing the warhead is on its way."

"Oh dear!" said Kelly. "We've got to stop them!"

"Yeah, but we need help, don't we?" asked Tom.

"Yeah, maybe, but let's wait a bit." Steve answered. "We have to be careful who we tell about this."

"Why is that?"

"We can't let the terrorists know we're onto them."

"I see,"

"And most of all, we need a plan."

Homeland

General Lee had watched as the two bombs were carefully moved from the design center to the shipping department. Each one was handled by a select group of four of his most trusted assistants. These young women had been selected for their skills, loyalties, devotion to country and most of all their independence. None of them had families. Lee had hired them directly out of the orphanage, at the age of eighteen.

The women handled the bombs with the greatest of care and saw that they were suspended in special crates that would protect them from any impact or jarring while in transit.

General Lee treated his women well. They had pleasant, heated, and air-conditioned quarters, furnished with expensive beds, linens, and elaborate baths. They dined apart from the other workers on excellent meals prepared by their own private chef.

Standing back, General Lee let them do their work, all the while keeping a keen eye on the proceedings. He would not relax until the cargo was safely loaded onto a train that would take it to the seaport.

Risky Offer

Meanwhile, Kim Jung Lo conferred with his trusted assistant, Hua Ming Du. Jung Lo drew him aside into a quiet space and spoke in low tones. "Ming Du, I must tell you, I have serious concerns about our safety, now that the project is completed."

"Why is that, sir? Aren't we going to be feted at an award ceremony?"

"Yes, tomorrow we will all be assembled, and given our medals."

"I shall look forward to that honor, won't you?"

"You've been told to have your personal things gathered, prepared to go home, right?"

"Well yes, haven't you?"

"Yes, that's true. But, how will you go home?"

"On the train, I suppose. It's the only way out of here."

"Yes, the train. But, have you noticed what direction it goes?"

"Not really. I've lost all sense of direction."

"I have not. The train goes north, Ming Du. North to the death camps, not south to the cities."

"Surely not!" Ming Du blanched.

Jung Lo merely nodded and glanced around to make sure no one was near. Had he made a terrible mistake, confiding in Hua? It was a calculated risk. If he took Ming Du with him that would increase the risk of being caught. But Hua was the only one who knew that Jung Lo was the last person to touch the computer control program. What if the tampering was discovered? Jung Lo weighed his options.

"And so what are you going to do?" asked Ming Du.

"I have a plan." Jung Lo's decision was made.

"What?"

"If you want to go with me, I will share it with you and no one else."

"Oh ... I don't know ..." Hua Ming Du hesitated.

Jung Lo drew in a deep breath. "Well then, Hua Ming Du, I must ask you to speak of this to no one else. Do you promise?"

Ming Du merely looked away. Alarmed, Jung Lo was already regretting his decision to invite Ming Du in on his escape plans. This could be a fatal error. Would he have to silence Ming Du? He grabbed Ming Du and hissed into his face. "Promise on your mother's grave or I will kill you! Do you hear?"

"I-I promise."

"You will not betray me," he spoke earnestly, placing one hand around Hua's throat.

Later that night, lying in his bunk, Jung Lo was wide awake at a few minutes before midnight listening to the snoring of his bunkmates and waiting for the sounds of footsteps as the guard neared their barracks for the last bed-check of the day. Jung Lo closed his eyes and feigned sleep as the flashlight quickly swept over the inert forms of his fellows.

There would not be another bed-check until almost dawn.

Jung Lo knew this was his only chance to avoid the so-called re-education camps. He rose and arranged his covers as best he could, hoping to fool the early morning guard. He grabbed a bundle from under his bed and crept out in stockinged feet. He would carry his shoes until he was safely away.

It was a cool clear night. The stars were out in splendid array but the moon had not risen as yet. Jung Lo could see well enough to make out the outlines of buildings. He would head generally north following the star, when he could. When he was discovered missing, the soldiers would expect him to go south and would most likely look for him in that direction. They would drive off in jeeps following the bumpy two-tracks. Jung Lo would avoid the dirt lanes. Instead he would sleep by day and walk by night in the creeks and through the fields and woods. He knew how to find food in the wilderness and build shelters.

Jung Lo had managed to assemble a few survival tools and hide them away. He had a knife, fish line and hooks, a crude trap, fire-starter, and a water bottle. Many of these things he had found tossed in the trash or crafted himself from simple materials he found lying around. He had two precious large plastic trash bags that would serve as ground covers or rain shelters. He had fashioned a backpack out of an old shirt.

Now that he was on his way, he was glad that Hua Ming Du had not come. Hua would have been a burden.

On the Trail

Jung Lo had watched when escapees were brought back to camp. So far as he knew, none had survived. Instead they were placed in the "sweat box" next to the walkway where everyone passed by on their way to the food line twice a day. The prisoner was locked in so that his head was outside the box. And so, for the first day or two he would cry out for water as others passed by. No one dared to look, much less help him. Sometimes the guards would gather to taunt and laugh at the poor man, eat, drink and place wagers on how many hours he would last.

No doubt it was fear of such a possible fate that kept Ming Du from joining the escape attempt.

Jung Lo had planned and prepared. He made friends with the dogs and would often visit them in the kennel to play. When possible, he brought them little treats, squirreled away from his own rations or the occasional rodent he caught. For the last day he had been preparing his shoes in readiness. They were soaked and rubbed with doggy urine and feces and hidden in a safe place.

After tiptoeing out of the barracks, Jung Lo retrieved the shoes and hung them around his neck. He headed East out of camp where he entered a small stream and began wading north. He would walk until dawn and then grab a thick tree limb hanging over the water. He would climb the tree and remain there until dark.

Just as expected, he did not hear the barking dogs until an hour after dawn. The lazy guards would not begin their search until they had breakfasted. Why hurry? They knew the dogs would find the escapee soon enough. Most times the victim had merely walked down the path leading south out of camp. Given the scent from the worker's bedclothes, the dogs could track him down within an hour or two, sometimes minutes.

In an attempt to foil the hounds, Jung Lo had sprinkled his bedclothes with urine from the female dog, hoping that would slow the tracking dogs down.

Shortly after dawn, Jung Lo had tied himself to an overhanging tree with a makeshift sling fashioned from clothesline rope. This allowed him to relax without fear of falling. In time he would learn to sleep this way, but for now there was no sleep. The distant baying of dogs kept him alert. For the first hour the sound had gradually gotten softer and stopped, telling Jung Lo that the dogs had gone south out of camp until the guards had called the dogs to jump into their jeep for the ride back to camp. Jung Lo knew they would be given a fresh scent and sent off to the south, down the creek bank, seeking a new scent to follow. This had happened so many times that a trail was beaten down the creek, making it fairly easy to follow the dogs on foot. Jung Lo was counting on the guards being too lazy to send the dogs north. This part of the creek bank had never been opened. It was almost impenetrable with thick brush and briars. So far as Jung Lo knew, no prisoner had ever been brought back from that area. Most of those clever enough to travel north had taken to the fields away from the creek. They were soon caught.

Soon he heard the dogs baying again and figured they had been sent off in a different direction. Would they come his way? Probably not until they had exhausted every other option. By then it would be nearing nightfall and Jung Lo would drop into the creek to continue his trek. He would not leave the creek until he was certain the dogs and guards had given up. After all, the mission was over. Surely there was no need for anyone to stay.

Army Orders

Indeed, Jung Lo was more correct than he realized, for that very morning Dear Leader had new orders for General Lee.

"Come in, come in, General," DL greeted Lee at the door in uncharacteristic fashion, clasping him on the shoulder. "Bring two coffees and biscuits," he ordered the lovely servant.

"Sit, please, General," he gestured toward an easy chair next to a beautifully carved and etched teak-wood coffee table. "I'm so pleased with the job you did, General. Nice work, very nice indeed!"

"Thank you, Dear Leader, but I cannot take all the credit. Without your direction and generous support, I am nothing."

"Nonsense," DL insisted, as the waitress came in carrying a large silver coffee service which she placed on the table. She unfolded an embroidered linen napkin and placed it across Lee's lap. Then she separated two exquisite china saucers and arranged the matching teacups in place. As she carefully poured from the silver coffee pot, her beautiful long hair cascaded around her body filling the air with the scent of jasmine. Lee was mesmerized by her graceful hands as she opened the silver tongs and lifted a delicate pastry from the serving tray onto a china plate and offered it to the bug-eyed general.

General Lee took the delicate plate in hand and set it on his lap, feeling very much out-of-place. Unsure how to balance the coffee and the plate at the same time, he waited for DL to offer some clue.

"Now, dear General Lee," DL began. "I want you to know how very pleased we are with the successful completion of your assignment." He ignored the sweet pastry and picked up the coffee. As he sipped, his eyes bored into General Lee over the rim.

Feeling like a bug-on-a-pin, Lee could barely breathe. A bead of sweat popped out on his brow. He froze in place, forcing his hands to remain still. Noticing DL's reference to 'we', he wondered who 'we' could be. Lee braced himself for the worst and clung to the thought, "I have the insurance policy."

"Relax, dear boy," DL chuckled, enjoying his power over this stupid little man. "Have some of this delicious pastry."

Instead Lee placed the pastry plate down on the table and picked up the coffee cup. He figured one or the other could be poisoned. The coffee seemed safer. Didn't he see the woman pour both cups from the same urn? But wait, the poison could have been in the cup already, couldn't it? Then he remembered she had turned the cup over. But, did she slip the poison in after she turned the cup over? Stomach churning, Lee placed the cup to his lips. Unable to take a sip, he set the cup back down. It jiggled a bit as he placed it on the saucer. Lee grasped his hands together in his lap to stop them from shaking.

DL set his coffee cup down, as well, and leaned back. "Now, General Lee, I presume you have the controllers with you, right?"

Lee nearly choked on his words. "Uh...N-no, sir"

DL frowned, "And why is that?"

"Well, uh, they are being examined, sir."

"Examined?"

"Uh, making sure they are completely ready, sir."

"Ready?"

"Yes, sir."

DL rose to his feet and summoned an armed guard.

The guard marched stiffly into the room and saluted DL. He stood at attention. Huge, at more than six feet tall and 250 pounds. He had two firearms buckled to his hips and wore ankle-high, spit-shined boots covered in white spats. His broad chest was adorned with many-colored strips representing his medals. His sleeve showed the stripes of Master Sergeant.

Lee turned pale.

DL returned the salute. "Master Sergeant, please escort General Lee to his quarters, after which you will accompany him to the design center or wherever he needs to go to fetch the two controllers. You will, then, bring the controllers directly here to me personally. Thank you. Please carry on."

"Yes sir!" said the guard.

Lee was so startled he froze in place. His mouth gaped open as he stared from the DL to the guard and back.

DL turned to Lee, "We can't take any chances with the control devices, now, can we, dear boy? I'm shocked that you would let them out of your sight. But, no matter, this fellow will make sure nothing happens to them, won't you, Sergeant?"

The guard reached for Lee's arm to help him out of the chair. Briskly they moved to leave the room.

Chapter 11

Senator McBride's Office

"Come in, Ms. Patterson and tell me what's going on in Honduras," said Senator Mike McBride.

"Thank you, sir," said Cynthia, as she took a seat beside his desk. "We've heard from our USA team in Honduras. Just now they have some startling news. They have done amazing things."

"How so?"

"Well, they managed to intercept an incoming message on the terrorist's email."

"You're kidding me!"

"Not at all, sir."

"How on earth did they do that?"

"Like I said, amazing. First I think you need to know the message."

"Sure, go ahead," said Mike, leaning forward in his executive desk chair, his curiosity killing him.

"The message said, 'Device is ready. Shipment going out today. Watch for arrival' and it was written in Korean."

"What device?"

"Well, the team believes it refers to a nuclear warhead."

"But ... how ...?"

"They think that the terrorists have figured out how to mount a nuclear warhead onto a medium range ballistic missile."

"That small?"

"Yes."

159

"But, our military doesn't believe …" he paused, shaking his head. "Wow, Miss Patterson. I've been behind the times."

"That is why they have set up the launch from Honduras, because their missile can reach the USA from there."

"And Honduras is considered a troubled country, too."

"Yes, that's true."

"So, does that explain why there is just one missile?"

"Yes, it only takes one," Cynthia sighed.

Mike paused, tapping a pencil on his desk, his brow furrowed. He swung around in his chair and gazed out the window. Turning back he asked, "What can we do to help?"

"Well, it seems obvious to me that we must try to stop the shipment. But, how? We have no idea how it is being shipped."

Mike considered that for a moment. "Remember the numbers and letters on the outside of that shipping container? You were going to look that up."

"Yes, I did that, sir. It gave the name of the shipping company, the port of departure and the port of entry."

"So, then, maybe we can get a line on any similar shipments."

"I'll try that, sir. But the problem I see is that this is a much smaller item and so they wouldn't put it in a shipping container, would they? Besides, wouldn't they want it faster?"

"Good questions."

"Maybe they sent it by air, you know."

"Well then we will have to cover that, too," Mike reasoned.

"Good plan, but would a commercial airline take a nuclear warhead?"

"Certainly not if they knew."

"Then, how ...?" Cynthia bit her lip. She stood to leave. "Sir, I sure hope we can stop it long before it gets to Honduras."

Back in her office, Cynthia went to work. The original container did indeed ship out of a large far-eastern port via a commercial shipping line. In New York it was transferred to a ship steaming to Miami, where it was transferred again to a smaller ship which stopped in several local ports in the Caribbean and Central America. It then passed through the Panama Canal and up the Western coastline before it returned to Miami. The entire round trip took several weeks.

The satellite's shipping method could not be traced. She spent half a day researching every known carrier. How did they do it? Her final call was to her husband, a Major in the Air Force.

"Hi Jake, is Sky available?"

"Hold on a second Cynthia, I'll see if he's through talking to the General."

Jake was Major Sky Eastman's aide. He knew that the only person who he could put through directly no matter who was there, or who Sky had on hold, was Sky's wife, Cynthia Patterson. Jake signaled Sky that his wife was waiting on line two. Their secret signal was two fingers patting on the heart.

"Thank you for the call, General," said Sky, "I'll get back to you, sir." He switched to line two. "Cynthia?"

"Hi darling, yes it's me, the pest, calling to interrupt your busy day."

"Thank you for interrupting me, sweetness. You saved me from a boring phone call."

"I'll not take long, honey. I just need to pick your brain about a problem I'm working on."

"You can pick my brain anytime you want, sweetie. Pick away."

"Here's a hypothetical problem. If you were a dictator shipping an atomic explosive from Korea to a small island off the northeast coast of Honduras, how would you do it?"

Sky gulped. What was his wife up to, now? "Hold on, Cynthia. I need to take this on a secure line. Hang up, I'll call you back." A normal person would have laughed at such a preposterous question. But Sky knew his wife wouldn't joke about this on a Pentagon line. He hung up abruptly and moved into a different room, one he knew was secure. He also knew that Cynthia would be moving into a secure room in the senator's offices if she wasn't already there. In seconds Sky called her on that number.

"Hello Darling," Cynthia answered. They would not use names on this line. It was secure, but … was anything secure nowadays?

"Okay, go ahead. Maybe you should give me a bit more background," said Sky.

Cynthia filled him in on everything she knew. "And so, I've checked every shipping line and air freight service," she finished.

"Well, the Air Force has a satellite that monitors Southeast Asia constantly. It broadcasts on a safe frequency that cannot be breached. Whenever a rocket is launched we know instantly the speed, direction and velocity."

"And what about Honduras?"

"None. We have four such satellites, but we only cover the countries that have the capability to launch intercontinental ballistic missiles. We aren't worried about

162

Central or South America, or our northern neighbor either, for that matter."

"So that explains why they chose Honduras."

"Yes, I'm afraid it does."

"Okay, so how are they moving the weapon?"

"You realize they have a Navy and an Air Force, don't you?" asked Sky.

"Why didn't I think of that?"

"Well, probably because all we see, when they hold their fancy military parades is their army. However, they have airplanes, ships and submarines, outdated, but useable. Mostly, they stay around the Homeland for defensive purposes, so you don't hear much about them."

"That's got to be it," said Cynthia.

Homeland—at the Project Site

General Lee smiled at his captor, congenially. "Here's my office, sir." He unlocked the door and held it open for the guard to step through.

"You first," said the guard motioning with his gun.

Lee stepped through the door and over to his desk. He reached for a drawer.

"Hold it right there!" commanded the guard.

Lee stopped mid-air. "I was only going to get out the drawings," he chuckled.

"What drawings?" The guard was suspicious of General Lee. Maybe there was a weapon in the desk drawer.

"It's the specifications for the controller device," said Lee reasonably.

The guard stepped back and took a firing stance. "Okay go ahead."

Lee pulled out some papers, spread them on his desk. He was taking a chance, hoping that the guard had no idea what a controller looked like or what it did. The papers were covered with intricate engineering drawings and long lines of indecipherable specifications. "See, here it is."

"Never mind that," said the guard, unwilling to admit his ignorance. "Just get the controllers for Dear Leader."

"Well, actually, these are just the plans," said Lee indicating the papers with a sweep of his hand.

The guard lowered the gun but stood his ground. "We need the controllers," he insisted

"Well, as I said, these are the specs. We have to build the controllers according to these specs."

"What!"

"You understand, we must take these instructions and build a device."

"Are you trying to tell me that you don't have the devices?"

"Well, not yet, but we will."

"Lies!" accused the guard as he took a menacing step forward.

Lee dropped into his desk chair and leaned away. "N-n-no, I swear...it's t-true." Lee shrank back and involuntarily shoved his hands under his thighs.

"You're lying! Bring me the controllers, now, or I'll start cutting off your fingers, one at a time." In one swift move, the burly guard holstered his gun, grabbed Lee's hand and slammed it on the desk. Suddenly a gleaming knife appeared, poised and ready for chopping fingers.

"N-no! Please!" Lee shrieked as the guard took a tiny slice. A bead of blood appeared. Lee knew exactly how this was done. He had done it many times, himself. He had to bargain with the man. "All right, I'll do it."

The knife cut deeper. Lee screamed.

The connecting door opened a crack. Lee's beautiful young assistant peeked in. Her mouth gaped open. "Sir?" she gasped, horrified at the scene.

Without letting go, the guard turned his head enough to see who it was. "Who's the cunt?" he asked Lee.

"She can get them for you, if I tell her," Lee saw a slight advantage.

Lee's blood was dribbling onto the papers.

The guard lifted his blade slightly. "Make it quick."

The woman closed the door, prepared to run. "Stop," yelled Lee. "I need you to go down to the storeroom."

On the other side of the closed door, the woman hesitated.

"Please, listen. He won't hurt you," Lee lied.

She turned and opened the door a tiny crack, prepared to run at any second. "I'm listening."

"Good," said Lee, thinking fast. "I need you to go down to the storeroom and bring me two of those green canisters—the ones on the far end."

"The empty ones?" she asked.

"No, the ones with the labels and the on/off switches."

"Uh..." she started to speak.

"Don't ask questions, just go. And hurry." Lee knew she was confused but she was a smart girl. He hoped this would work. Would she understand his ruse and slap on some fake labels?

Five minutes passed. Beads of sweat ran down his forehead. The blood continued to ooze. "Please can I wrap up this wound? The blood is ruining our drawings."

The guard nodded and lifted the knife.

Lee pulled a white handkerchief out of his pocket and wrapped it tightly around the wound. There was nothing he could do about the blood drippings. He sat silently waiting.

Finally a tiny knock came on the door. "Bring it here," commanded the guard. Saying nothing, the woman merely shoved a green canister through the door with one hand while tightly grasping the door handle.

"Get it," the guard took one step back and pulled the gun, waving Lee toward the door.

Lee took the canister from her hand. "Thank you," he said. "Did you bring them both?"

"Yes, sir," she toed the second one through the door.

Lee was relieved to see a fresh label stuck on each one. "You did well, my dear. And now I want you and the other women to take the dogs and go to the resort, immediately."

"All of us?" she asked.

"Yes, all of you go. You'll be safe there. The company will pay your expenses. Take some money from the petty cash. Now go!" The door closed and he could hear her running away. He turned toward the guard. "Shall we take these to DL?"

Further Orders

General Lee presented the fake controllers to Dear Leader with a flourish. "Thank you for your patience, sir. We have your controllers right here."

DL took one of the green canisters and turned it all around.

"Please, sir, may I say, please be careful with this delicate device? I wouldn't jiggle it or move it around. And, whatever you do, never touch the switch or you will set off the bomb."

DL jumped away so fast he let go of the device.

General Lee caught it just in time. He drew in a deep breath and let it out. "Whew," he carried the device over to a corner table and set it down as if it was a newborn baby. "That could have been a disaster, sir. We don't want to set off the bomb while it is still on the submarine."

DL's eyes were like saucers. Lee offered him the second fake-controller canister. DL took a step back and threw up his hands in defense. "No-no. Just put it over there with the other one."

"All right, sir," said Lee smiling to himself as he gently carried the second canister over to the corner table. "Shall we leave them right here for now?" he asked.

DL nodded and moved to his desk chair in an effort to get as far away as possible from the things.

Lee was on a roll. "Do you have a marker pen in your desk, sir?" he asked, innocently.

DL opened his center desk drawer and extracted a pen, offering it to Lee.

"Thank you," said Lee. "I did not have time to mark these for you. Give me just one minute and I'll take care of that." He moved over to the corner table, picked up one canister and pretended to read the label. "Ah, this one controls the weapon for satellite number one. I'll put a large number one on the label." He began writing. "Also, notice I wrote the word Island. This controls the bomb on the Island. The other one is number two, Honduras. I'll write that out. This one will come from the mainland Honduras. Not that it matters all that much," he chuckled. "Either bomb will wipe out North America. So don't worry about it if you get them confused. One will do the job and then you'll have the backup, in case you need it." Lee held up the canister. "See this button here? It's very simple.

You push that button and the bomb goes off. Understand?"

DL nodded.

"Again, just a word of caution. Be careful that you don't accidentally bump or jar the device. Maybe you'd better put them away for now. Can I do that for you?"

"No, Lee, just leave them alone. I want them close where I can watch them myself."

"Makes sense, sir. Will that be all?"

"Uh, no, General. I have another assignment in mind for you. Please sit down."

For the first time, Lee realized that the guard had disappeared. Now that DL no longer needed him, was this going to be Lee's final farewell to earth? Lee hoped he would not embarrass himself when it came.

"Now, General Lee, I have been very impressed with your ability to command. You completed a difficult assignment and fulfilled your mission admirably. And so I have decided to reward you with a new mission," DL announced.

Lee braced himself and tried to appear dignified.

"You will take command of the invading armies."

Lee had been so focused on the worst that could happen, he did not hear DL's words. Instead he closed his eyes and held his breath.

"General Lee?" said DL. "Are you asleep?"

"What? Uh...eh...uh," Lee sputtered and opened his eyes.

DL was smiling. Lee stared in disbelief. He had never seen DL smile.

"Did you hear me, General? You will take command of the invading armies."

Completely dumbfounded, Lee managed to ask, "An invasion? Where?"

"North America, of course. You will take two armies and everything you need to take over Washington and New York to start. Everything they have will be wiped out—melted," DL chuckled, rubbing his hands together in glee.

Lee was stunned.

"You will take your crack team of scientists with their equipment, as well as your most trusted soldiers in the first wave. Your entire occupying army will soon follow, one-million strong, with everything you will need to take control of the United States and Canada."

"Well, uh, yes, sir, thank you, sir," said Lee, nearly tongue-tied.

Chapter 12

Major Sky Eastman Goes to Work

After talking with Cynthia, Sky went to work immediately. She had told him about the amazing and scary way her team USA had gotten pictures of the terrorists and their portable launch platform. If he was going to do anything to help, he needed to get a look at that launcher. This would tell him whether such a launch was even feasible. But where were the pictures?

Sky opened a call on the Pentagon's interconnected private communicator. Photo-analyst James Petrie's picture faded in. "Hello, sir, how may I help you?"

"I'm looking for pictures that came in from our spy-plane over Central America, three days ago. They show some people working in a wooded section of an island off the northeast coast of Honduras. Do you have them?"

"Please hold on one minute while I check."

"Sure, go ahead," said Sky as if he had all the time in the world. It was early in the day—not time to get frustrated.

"Ah, yes, sir, we have those."

"All right, can you send them over to my computer?" asked Sky.

"Let me check on that, sir. I may need to send them by courier. In the meantime, you might want to check with Mitch Mucurie over at CIA. They requested the pictures that first day. Could be they are doing some analysis."

"I take it the Air Force is not doing anything with them," observed Sky.

"Let me check here ...uh...no, sir. The request for the flyover came from the CIA and so we sent them the pictures."

"Thank you very much, James. Just send those pictures over to me and I'll take it from here."

Sky buzzed Jake, his aide.

"Yes, sir."

"Jake would you please get Mitch Mucurie, over at CIA, on the line for me?"

"Certainly, sir. One moment."

It took some minutes for the CIA operator to attempt to locate a certain Mitch Mucurie. She came back on the line, "Sir, can you spell that for me?"

Jake had to guess at the spelling. No one seemed to have heard of him.

"I cannot find an employee by either one of those spellings, sir. Could be he is a new employee. It takes a while before our files are updated."

"Okay, thanks, I'll have to call you back," said Jake. He buzzed his boss. "Sir, CIA doesn't seem to know a Mitch Mucurie. I tried different spellings."

Sky blew out a deep calming breath. "Um, well, try connecting to CIA's photo-incoming desk for Central America, and see if they ever heard of Mitch Mucurie. I'll hold." He listened while Jake went through the operator again.

A voice answered, "Central America photo incoming, Mucurie speaking."

"Mitch Mucurie?" asked Sky, feeling a bit relieved. Maybe he solved one mystery today and it wasn't even lunchtime.

"Yes, Mitch Mucurie here, how may I help you?"

"Uh, well, this is Major Sky Eastman, US Air Force. I'm calling regarding photos that CIA requested three days ago."

"Three days?"

"Yes."

"Uhm, well, we get a lot of photos here. Can you tell me anything more about them? That might narrow my search, sir."

"Oh yes, of course. These were taken of an island off the northeast coast of Honduras. The flyover request was made by the CIA operative there. His name is Steve Spalding. I don't have his number."

"Yes, sir, I have those right here in my pending file. We are waiting for Special Agent Spalding to submit his written report. Let me open the folder and see what we have so far. Um, yes, the Air Force pictures are all here, as well as the pictures taken by the CIA satellite. That is all we have, so far."

"You're new there, I gather," surmised Sky.

"Yes, sir, my three-month's anniversary is coming up this Friday."

"I had a little trouble finding you, Mucurie."

"Sorry, sir. This is a big place. It takes a while for things to get posted, I suppose."

"And so, the photo file has not been analyzed as yet. Is that correct?"

"No, sir. I'm trained to only do the intake. Analysis is the next paygrade up. I hope to make that by next year."

"I see," said Sky, tapping his fingernails in frustration. "Well, then, can you send me the satellite photos?"

"That isn't my department, sir."

"Do you know where I should go to get that done?"

"Actually, no, sir, I'm new here."

"Thank you, Mucurie. That will be all for now." Sky hung up quickly before he started yelling at the man. "Did you hear that, Jake?"

"Unbelievable, sir!" said Jake.

"How on earth did we ever win World War II?"

"I don't know, sir."

"Well, the Axis bureaucracy must have been worse."

"Yeah, I think you're right. It had to be that."

Sky sighed, "Well, Jake I promised to get back to the general. I guess he's got some new assignment for me."

"I can get him on the communicator for you, sir."

"Okay, Jake, thanks." Sky rose to pace the floor, stroking his chin, tapping his lip with one finger and shaking his head. Five minutes later, his communicator signaled.

The general was on the line. He wasted no time getting right to the point. "So, Eastman, let's have your decision."

"I'll do it, General," said Sky. "When can I start?"

"Just as soon as I get your upgrade to a full-bird Colonel," the general remarked. "Can't put a measly Major in charge of a new branch of the service, you know."

"Thank you, sir, but that won't be necessary, will it?"

"Actually yes it is. It wouldn't do to have a Major giving orders to a Colonel, would it?"

"I see your point, but I want to get started as soon as possible. Just give me a bunch of civilian techies and young 2nd Lieutenants. We have a crisis already."

"Fine, go for it. Colonel Eastman. It's your baby."

Kim Jung Lo Escapes Capture

General Lee wasted no time leaving DL's clutches and returning to the project. Although life had long ago stolen nearly every emotion from his mind and body, in a small way he was relieved of the necessity to send his crack team of scientists to the death camp. That would have been such a waste of talent. Instead they would pack up everything and move off to the troop carriers that would steam across the oceans toward North America. Lee was also relieved to see that the women had escaped with the dogs. If he had ever allowed himself to feel, it would have been a fondness for his loyal women. At this point he could not disclose the plans to anyone and so it was fortunate his guards were still there to enforce discipline.

Meanwhile, his chief scientist, Kim Jung Lo, was still missing. This was unfortunate, but nothing could be done about it. The dogs were gone, and Lee could not afford to take more time in the search. He turned his attention to preparations for departure. They would have everything they needed to power up their lab and advanced communications equipment as well as the necessary weapons and provisions to equip a small army.

The USA Team Has a Plan

Sharon and Special Agent Steve Spalding sat close together on the loveseat, while the others perched on chairs and cushions. The group were discussing their lack of progress on intercepting the shipment. "Cynthia has more search tools available than we do, but she has drawn a blank, too." Sharon shook her head. "I'm out of ideas. What does anyone else think?"

They contemplated each other. Finally Aggie spoke, "Well, if we can't intercept the shipment, it seems to me that all we can do is stakeout the terrorists. What do you all think?" She looked around for an answer.

Tom spoke. "I guess I can take the first shift. I've had plenty of experience at staying awake during long boring stakeouts."

"I can go next," Kelly volunteered.

"All right, folks," said Steve. "We'll all take a turn. Everyone else can keep their phones open and batteries charged."

Tom moved to the kitchen, "I'll start the coffeemaker."

"How long before I relieve you, Tom?" asked Kelly.

"Well, I can talk to you on the phone, but I'd say no more than five hours. There are five of us, so that means we each take a shift once a day. Does that sound about right, Steve?"

"Sounds good to me," said Steve. "Meanwhile, we've still got their cell phone. Maybe another message will come in. And I have my contacts watching their leader's movements. I'll know if she leaves her hotel."

Bombs Travel

The bombs had successfully traveled several legs of the journey. The first leg by submarine took the longest until they met up with a much faster cruiser. The third leg was by one of the Homeland's fighter jets, which landed on one of the Caribbean Islands where an innocent-looking pleasure boat took over for the final leg.

Colonel Rhee Su-jin nervously awaited the arrival of her very special shipment. Her orders were to meet the pleasure boat at midnight offshore in a sheltered cove near one of the hundreds of keys that dotted this area of the ocean. How would she manage this without alerting the authorities? She had no idea how to operate a boat. Could she hire one of the charter boats that tied up at the pier? She remembered the handsome American who had invited her on a sailboat. Maybe she could figure out a way to get his help. Her mind went around and around trying to come up with some plausible lie. She wasn't even sure where to find this man. He seemed to just show up out of nowhere. No, that simply would not work. Getting a stranger involved was too risky.

Su-jin hated to admit defeat, but she had to do it. She messaged the Homeland. "Change plan. Cannot do meeting as instructed. Have pleasure boat tied up at the island pier and await instructions."

Col. Rhee Su-jin Meets the Delivery

Sometimes the direct approach works best. Rather than skulk around at midnight, Su-jin waited until lunchtime, put on a casual outfit, her fancy hat and shoes, carried a large tote-bag and hired a taxi to take her to the pier. Now she had the problem of figuring out which boat had her package. How in heck would she find it among the hundreds of luxury yachts bobbing quietly in the sheltered cove? All were tied up to expensive rented positions along dozens of arms on the four different piers in the harbor. She looked around in amazement. Where on earth did people get all this money? Millions and millions in value stretched out before her eyes. Maybe this wasn't a good idea, after all.

Did she dare send another message? She could do it, if she hadn't misplaced the cell phone. It was inconvenient, but probably safer, having to use the computer for messages.

Feeling hopeless, Su-jin decided she must return to the lab. She hired another taxi and directed the driver to The

Pigsty Gift Shop. Pulling up in front, the driver parked, got out and opened her door. Without a word, she offered him a bill and waved off the change.

"Gracias, senorita," said the driver with a broad smile.

Su-jin unlocked the front door, entered the shop, pulled out a stool from behind the cash register and sat, morosely leaning her elbows on the counter, chin on her fists. As she gazed out the window, she noticed a delivery truck pull into the empty space out front. A uniformed man jumped out and opened the back of the truck. He walked into the shop and called, "Delivery for The Pigsty Gift Shop. Where would you like it?"

Readies Launch

Su-jin was almost tempted to believe in miracles. Here was the device delivered to her door. She had the delivery truck driver go around to the back of the shop and unload it into the shipping container. Now she had everything she needed for a successful mission—the launcher, missile, satellite and explosive device. She was prepared. It was just a matter of waiting for orders to launch. Unfortunately for her, she did not know that other eyes were watching and waiting.

How Will USA Team Thwart Island Launch?

Team USA now knew that the terrorists had received some sort of delivery. But was it the final piece of their puzzle or simply another order for the gift shop? They certainly had no way of knowing without breaking into the container again or intercepting a crucial message. Even if they figured it out, what could they do anyway? This was a foreign country. They couldn't just step in and arrest people without bringing in the local police, which was out of the question.

Team USA consulted together by cell phone. Tom had just reported sighting the delivery. "Well, Steve, what do we do now?" asked Kelly. "You're CIA. You tell us."

Chapter 13

Launch Day

Colonel Rhee Su-jin was in her glory, shouting orders to a well-schooled team. They scurried around putting up the launcher in record-breaking time. The missile gleamed on the launch-pad, the satellite with its payload not yet affixed to the nose.

Well hidden in the surrounding woods, team USA watched and waited. Would their plan work?

Special Agent Steve Spalding's tiny USA team were willing to attack even though they were outnumbered. But Steve would not allow civilians to take such a risk. This was his job.

Dressed in camouflage and armed to the teeth, Steve sneaked, unnoticed, up behind Colonel Rhee. He waited until her team was distracted, busy unloading a heavy crate of equipment. In one practiced move he grabbed her head and put all his might into a near-fatal twist of her neck. Su-jin melted into his arms. Steve swiftly dragged her out of sight where his team was waiting to tie her up. Steve grabbed her communicator and began issuing orders to her crew. Meanwhile, a military helicopter from nearby Guantanamo Bay appeared in the sky above. The terrorists obeyed Steve's command to lay down their arms and move away from the landing site, as the chopper settled amid whirring blades. A combined team of CIA and special forces took the terrorists into custody without firing a shot.

Team USA gathered in the clearing to watch the chopper take off. They stood around the launch platform and gazed into the sky as the chopper faded into a tiny spot and disappeared.

"Good job, Steve," said Sharon, giving him a one-armed hug. "So what do you think?"

He grinned and hugged her back. "I think we need to disarm this thing somehow."

Space Force Detects Second Launch

USA Space Force's brand-new commander, Colonel Sky Eastman addressed his Space Force via his newly integrated communication system. This was the very latest, state-of-the-future device. It was versatile and tiny enough to fit into a human ear. It could also be molded into eye-glass frames, goggles or headgear or sit on a desk. Using this device Sky could talk to his entire force, or to any single one or group of individuals. The system instantly obeyed his spoken commands.

"Hello, people," said Sky. "Here's the situation. Our USA team in Honduras was successful in taking out the terrorists before they were able to effectively launch their rocket." His announcement was greeted by cheers. "But, we have big trouble. We did not know they had a backup system. Our spy plane has detected a launch from another position in Honduras. We are watching it now. The trajectory indicates it's heading over the central United States. Expected pass-over in ten minutes and counting." Sky heard a collective gasp. "We need to prepare to intercept immediately! North Dakota, please standby to launch missile defense."

"North Dakota, standing by for orders, sir!"

Sky paused, watching the computer read-out of the progress of the missile. "Hold on, North Dakota, we see separation!...What th' hell?...Oh dear God."

Sky watched in horror as a tiny satellite separated from the missile and seemed to continue while the missile fell away toward re-entry.

Dear Leader and General Lee's Progress

Back in the Homeland, Dear Leader stayed connected to General Lee, relishing every minute of Lee's reports.

Onboard the troop carrier, Lee was watching progress of the missile on the temporary computer lab and communication hub they had already set up. He needed to command his forces from here as they steamed across the ocean toward North America.

"Congratulations, DL," Lee boasted. "I believe the satellite is settling into the planned north-south LEO--Low Earth Orbit–sir. You should activate the bomb, immediately."

"Very good," said DL in a somewhat congratulatory tone. Praise by Dear Leader was so rare as to be almost non-existent. "And so, tell me again, how long it will take the satellite to make one orbit?"

"About ninety minutes, sir. You should activate the bomb now, sir, before it sails out of range." Lee knew that DL's controller was a fake, but Lee had the real one in his hand. He could activate the bomb as soon as DL pushed his fake button. DL would never know the difference.

DL picked up both controllers. Unsure what to do he studied the marks.

"Go ahead, sir," said Lee "You need to activate the bomb before it moves away from the United States."

"Which one of these damn things do I use?" DL's voice rose.

"The one that says 2, sir, just push the button on 2," he said, forcing his voice to remain calm, unlike his nerves which were shrieking, *Push the damned button!*

"I don't see a 2!" DL yelled louder as if that would make his voice heard around the world.

"It's right there on the label. Turn it around. See the label where I marked it with a 2?"

"No I don't see any damned label!" DL screamed.

Lee was perplexed. What happened to the label? "Try the other controller," he suggested. It did not matter which button DL pushed. Both of them were fake. By now the satellite was passing over Canada. The window of opportunity was rapidly closing. General Lee could not be sure the satellite would live to make a second orbit. This whole thing was an experiment.

DL picked up the other controller and turned it around, looking for the label.

"Maybe the labels came off," suggested Lee.

"Oh son-of-a-bitch, here they are," DL cursed, bending over to pick up the papers from the floor. "Your damn cheap labels fell off. Now what the hell do I do?"

The satellite had already passed over Canada and was out of range of North America.

"We'll have to wait, now, for the next orbit," Lee sighed, trying to sound confident. He knew that the drag on the satellite would gradually slow it down until it eventually fell to earth. Hopefully they would get a second orbit. If only DL had allowed the test flights this would not have happened.

"Well, will it come back to this exact same spot each time?" asked DL.

"No it will gradually move to the west with each orbit. We'll know about that for certain, when it happens, but, yes, I think so. That is the plan."

"Well, what if it doesn't?"

"Our engineers will analyze the trajectory and make tiny adjustments as needed, with the onboard thrusters," he lied. The tiny size of the satellite did not allow for a second burst from the thrusters. All the fuel had been spent just getting it into orbit. If the satellite did not fall out of the sky, it would circle the earth, moving one and a half time zones to the west, as the earth spun on its axis, west to east. By the time the satellite made another orbit, it would already be over the Western United States. A third orbit would put it out over the Pacific Ocean. It would be twenty-four hours before it returned to the US. If DL didn't act, Lee would simply have to set off the bomb himself and hope it shut down New York and Washington DC, in addition to the West Coast. He knew that the low altitude of the satellite limited the visual field, but there was no sense trying to explain that to DL, who thought his mere wish was a command. What a mess!

DL was impressed. "Very good," he allowed himself one more kind word. "How soon will you have your invasion forces in place, General?"

"We're steaming across the Atlantic right now, sir. We'll be ready," he lied again. No need to tell DL the truth. It would be at least twenty-four hours before they made it to North America. That was okay. He wanted time for the radiation fall-out to decay, somewhat, before he landed his forces. They had hazmat suits, but still, he knew those suits were never one-hundred percent effective. His men needed to be able to fight, if necessary to overcome any spotty resistance.

"Atlantic? But isn't that the long way around?" asked DL, unwilling to show his ignorance of geography.

"Yes, sir, we came by way of the Suez Canal, following your instructions to take New York and Washington D.C., first." Good grief, did he have to remind DL of everything?

"Yes, but wouldn't it be faster to travel east across the Pacific?"

Lee suddenly realized that DL did not know that the target cities were on the East Coast of the US, rather than the West. How could he say this without insulting Dear Leader? "Well, sir, we thought it unwise to go through the Panama Canal because of the real danger of being detected by United States spies. We thought we could take them by surprise. They would not be expecting an invasion from the east."

"Um," said DL, pretending he understood east-west, west-east. By now he was thoroughly confused. "Well, then, General, as soon you have your forces in position, we'll explode the weapon."

"We'll be ready, sir, but if I may suggest, sir, we need to explode the bomb on the next orbit."

"All right, we'll talk later," said DL, ending the conversation.

Washington Panics

Having rushed to the situation room, President Gerard Bigelow gazed around the table at his tight-lipped advisors whose stricken eyes seemed to be glued to the TV set. Among them sat the vice-president, the chief of staff, two military advisors, the Secretary of Defense, the

Secretary of State and a couple other cabinet members who happened to be in his office at the time. Mrs. Beth Terry, his faithful personal secretary sat next to him, poised and ready to take notes by hand, since no recorder was allowed in this room.

Pentagon Reacts

Across town at the Pentagon, the five-star general who was chairman of the Joint Chiefs was poised to act on President Bigelow's orders. Instead, ignoring the chain-of-command, Bigelow spoke directly to his new Space Force Commander, Colonel Sky Eastman.

"What do we know, Commander?" asked Bigelow.

"We have a hostile satellite in a polar orbit progressing over North America, at approximately 450 to 500 kilometers of altitude, sir," Sky answered. "The ballistic missile has gone on, sir, and is no longer a threat."

"Go on, Eastman, explain. How in hell did this happen?"

"The missile was launched from a place on mainland Honduras, sir. Apparently, it had a payload of a small satellite."

"So why didn't we knock it out?"

"It did not come within range of NORAD'S defense missiles, sir, until after the satellite separated and the booster missile continued losing altitude. It has since disintegrated in the atmosphere and the remaining debris has fallen into Beaufort Sea."

"Honduras, you say? I'll get their president on the phone immediately."

"We have reason to believe that Honduras did not know about this, sir."

"What reason?"

"We now believe that this was one of two rockets secretly set up by a hostile power without the Honduras authorities' knowledge. The other rocket was stopped on the ground by our forces."

"Well, you were half right," said Bigelow sarcastically. "So get off your butts and do something to get rid of this satellite. And there damn well better not be a third rocket aimed at us."

Sky blanched from the rebuke. "Yes, sir," he said, "We are examining the satellite, sir."

"Examining? What's to examine?"

"We need to determine the capability of the device, sir."

"What th' hell does that mean?"

"At this point we do not know the purpose or exactly what is on board the satellite. It could be harmless, sir."

Cynthia's USA Team Examines Launch Platform

"Let me help you with that," said Tom as he hurried to give Steve a boost.

Steve already had one foot on the launch platform, ready to scramble up to the top.

"Are you crazy?" screamed Sharon, running forward. "You can't climb up that thing!"

Steve removed his foot. "Well, do you have a better idea?"

Hands on hips, she craned her neck to look up. "All I know is you can't do this. We've got to be careful. That thing could collapse. It could explode!"

Having joined them, the rest of the team pitched in. "Don't be a hero," said Agatha.

"Let's talk about this," urged Kelly.

"Alright, talk," said Steve.

"Um, well, all the terrorists are gone, now, aren't they? We should be able to get into their lab without any trouble. I mean, maybe we can learn something from their computers," suggested Kelly.

"Or, their cell phone," Sharon added.

Meanwhile, in Washington, Cynthia was urgently trying to contact Sharon and Agatha.

Agatha pulled out her cell phone and thumbed onto her message. "Hold on, people," she said. "Cynthia sent a message." She read it aloud, "Space Force needs more info regarding satellite. Can you investigate?"

"Is that all she said?" asked Steve.

"Yes, it is."

"What more can we tell them?" asked Tom.

Chapter 14

Sky Attempts Dangerous Rendezvous

In a restricted area of the Utah desert an enormous set of doors began to rumble open, exposing a gigantic hanger. Men and women swarmed over a special cargo airplane like a colony of ants. They were uniformly dressed in gleaming white coveralls with a spanking new insignia embroidered on the front and 'US Space Force' stamped across their backs in large letters.

For months they had been secretly practicing and preparing for their first real space mission. Housed inside the cargo airplane's cavernous space was a three-billion-dollar space-transporter and its crew of three well-trained, but excited, astronauts. From time to time their strange looking vehicle with its unusual flight characteristics was mistaken for a UFO.

Slowly Space Force One with its precious cargo moved out of the hanger and rested on the tarmac while the pilot and crew went through the pre-flight checklist. It was a clear and dry day, perfect for flying. The pilot, Major Jim Hughes, moved his guppy-like airplane onto the runway and spoke directly with his commander, Colonel Sky Eastman. "Space Force One ready for takeoff, sir."

"Cleared for take-off, Space Force One, God-speed," said Sky. "Ascend to thirty thousand feet and launch."

"Roger, thirty thousand and launch," said Hughes.

At twenty-five thousand feet, Jim pressed a control that gradually opened the large rear cargo door revealing an amazing space-age vehicle secured on a track inside the yawning hold. Three astronauts inside the space-transporter went through their final preparations and announced "Space Force Two hummingbird is ready to fly, Jim. Let 'er go."

Jim ascended to 30,000 and put the airplane in a nose-up position, announcing, "Ten seconds to launch," he began the countdown.

"Three, two, one ... release."

The space-transporter slid along its tracks out the back of the airplane and dropped away into the sky. Jim and his crew held their breath, watching for the ignition fire to spew out of the rocket-ship.

"There it goes," said Jim, with some relief. "Hummingbird is on its way." He had no idea where they were going or what was their mission. His job was to safely return Space Force One to base.

Once the space-transporter, Space Force Two, also described fondly as the hummingbird, was underway, mission commander, Andrew "Flash" Gordon reported to Colonel Eastman, "The Space Force hummingbird is safely underway, sir."

Only now was Sky able to give them their orders. "You have a dangerous and top-secret mission, Flash," he began. "You are to prepare to rendezvous with an enemy satellite as it approaches western Nevada, from the south, flying over Baja, California at approximately 115 degrees longitude, and await further orders." Sky gave Gordon the coordinates and altitude of the satellite, the estimated orbit and time of arrival. "As soon as it passes the southern hemisphere, we'll give you the exact coordinates. Meanwhile, you can just park over Utah and wait. Take a nap or something."

Flash laughed, "Roger, nap time."

The Launchpad and the Lab

Special Agent Steve Spalding divided his team. "Kelly, why don't you and Sharon, go to the terrorist's lab and see if you can break into their computers? Learn everything you can about the purpose of this missile. Okay?"

Kelly nodded.

Steve continued, "Tom, Aggie and I will try to figure out how this blasted thing works. Maybe we can deactivate it and take it apart."

Sharon said, "Okay, I'll go with Kelly on one condition."

Steve looked at her. "Oh so now we have conditions, huh?"

"Yeah, I do. That is you don't do anything stupid, again, like trying to climb this thing. And be careful, darn it! That crazy missile could explode, you know."

Steve gave her a rueful grin. "Sweetheart, I didn't know you cared."

"Pfft, don't flatter yourself," Sharon turned on her heel and swished off.

"Keep your phone on," Steve called after her.

Space Force Two –Hummingbird

Flash programmed the coordinates and altitude into his computers that would bring him within sight of the satellite. As they neared the area, he fired small retrorockets to hover his spacecraft into the desired circle. Obtaining a visual sighting of such a small object would be difficult. However, he hoped to locate it with his onboard radar and then bring his hummingbird close enough to obtain pictures. Whether or not they would try to do more remained to be seen.

Team USA, the Crate and Box

Steve and Tom stood scratching their heads looking over the missile as if they might figure out what to do. Meanwhile, Aggie called, "Over here, guys." They moved over to see what she was doing. Aggie was poking around at a crate. "What do you suppose is in here?" she asked.

"Let's pry it open and see," said Tom, moving around the thing to look at it. "You got a crowbar handy?" he asked.

"Not right here," said Steve as he tried to grab ahold of a corner and pry it open.

"I'll check around and see if I can find a tool," said Tom. "Maybe a rock will work." He started walking back and forth on the site, examining the brush.

Aggie stood aside and watched. When Steve left to help Tom, she started patting the crate with sensitive fingers. Soon she found a lever that was cleverly concealed. She pulled and twisted the lever which allowed a small panel to shift. Behind the panel was a flat handle which could be lifted and turned. Slowly she pulled on the handle which opened to one side of the crate. Inside the crate was a stiff cardboard box surrounded by a layer of padding. Just as Aggie began to carefully remove the padding the men came back. "Couldn't find a tool," said Steve. "Oh, I see you've got it open," said Tom.

"Yeah, I just stumbled across a secret panel. Lucky break," said Aggie, modestly.

"Can we help?" asked Tom.

"Well, I'm just removing this insulating material," said Aggie. "The way this is packed, it could be an important piece, don't you think?"

"Um, yeah," Steve mused, looking on.

Aggie continued carefully peeling away the packing until she exposed the cardboard box. She was able to slip the crate away from the box. Steve came up and opened the top of the box with his knife. Inside the box was another Styrofoam container. "Oh, what's this?" she wondered out loud. Both men leaned over, staring at the thing.

"Looks important," said Tom.

"Yeah," Steve agreed.

"They were expecting a weapon, weren't they?"

Steve nodded and sucked in through his teeth.

"That crate looks exactly like the delivery I saw," said Tom.

"Oh-oh," said Aggie, taking an involuntary step back.

The men glanced at her. Tom bit his lip, "How can we tell for sure?"

"Cynthia wrote that the Space Force needs info."

"Did she say why?" asked Steve.

"No. That's all she said. Shall I write back?"

"Maybe you'd better call her," Tom suggested.

"Good idea," said Aggie as she picked up her phone and selected Cynthia's number.

Cynthia answered, "Heh, Aggie, boy-oh-boy am I happy to hear from you!"

"Hi Cynthia. How are things up there?"

"We're okay but worried, with all that's going on."

"Listen, I got your message and we're wondering just how far we should go in taking this thing apart."

"Oh my gosh! Where are you?"

"Well we're out here at the launch site. I don't know how much you know."

"Nobody tells me anything!"

"Oh well, then, we caught the terrorists putting up a launch platform to set off a ballistic missile. Steve took out the leader and the rest surrendered when the helicopter came in."

"Helicopter! What helicopter?"

"Steve called in a chopper from the base at Guantanamo. A bunch of mean-looking military guys swooped in like a flock of avenging angels, rounded up the terrorists and took off into the clouds. That's the last I saw of them."

"Well, I heard the prisoners were being questioned at the garrison at Guantanamo. That much I know," said Cynthia, helpfully.

"Good to know."

"So what are you doing, now?"

"Well, Sharon and Kelly went downtown to see if they can learn anything from the computers. Tom, Steve and I are here trying to figure out what the payload was supposed to do. We found this crate and started opening it."

"I see."

"So here's our question. Should we go ahead and finish opening this crate? It looks pretty important."

"Did you see what the missile looked like?"

"Well, yeah, but it's just a ballistic missile as far as we can see. It could have been meant to carry anything, you know, anywhere from a weather satellite to a bomb on board, for all we know."

"Listen, Aggie, hold on a minute. I'm going to try and patch you through to Sky. I'm sure he will want to know, first-hand, what you have found. And also, he may be able to fill you in on what is happening here."

"Okay."

"Hold on a minute," said Cynthia as she punched into Sky's communicator. The back of his head appeared on the screen. Obviously he was talking to someone. "Yes, sir, Mr. President. We're on top of it …. Yes, sir, we anticipate intercept within the hour … No, sir … yes, sir … I can assure you our best astronauts are on the mission … No problem, sir, I'll get back to you just as soon as we know anything ... Goodbye for now, Mr. President."

Sky took a big breath and blew it out. He swung around to face Cynthia. "Hi, doll-face. You're looking beautiful, as always," he said.

"Hey, kiddo, I have someone on the line who wants to talk to you," said Cynthia. "Go ahead Aggie."

"Hello Commander. Listen we're down here at the island launch site, looking at this setup. C wrote us that you wanted info on the device."

"Indeed we do! Are you saying that the combatants are not around?"

"They are gone, sir, under arrest. The people down at Gitmo may be able to squeeze something out of them, and we'll do what we can here."

"Well, I've seen the great photos you took of the launch platform, but I haven't seen the missile or the payload," said Sky.

"Well, hold on, I can get a shot of the missile right here." She turned her camera toward the missile. "Can you see that, sir?"

Sky let out a low whistle. "Perfect view of the base. Can you slowly raise it?"

"Sure, how's this?" Aggie slowly scanned the entire height of the missile.

"Can you zoom in on the nose for me?"

Aggie directed the camera at the point of the missile and zoomed in.

194

"That's perfect," said Sky, "but I don't see anything attached to the nose, do you?"

Aggie looked at Steve in question. He shook his head. "There's nothing up there."

"Nothing there, commander," said Aggie. "Were you expecting something?"

"We're chasing a satellite right now that we think came from the same outfit. It's a mini-satellite capable of carrying a small payload. It's the possibility of a payload that has us worried."

"Are you saying there was another missile?"

"Yes."

"Two of them?"

"Well, just the one, now. We think the twin is there on the ground with you. That's why we need you to find the payload, if there is one. We're thinking there may have been two alike. If so, yours can tell us more about what we are up against here."

"But, we don't see anything," said Steve.

"Wait a minute," said Aggie. "What about the box?"

"What box?" asked Sky. "Show me the box."

Aggie directed her camera toward the crate and the box and described what he was seeing.

"That could be it," said Sky. "You had better clear out of there and I will send some people down to take care of whatever is in that box. It could be very dangerous."

"But, you need to know now, don't you?" insisted Aggie.

"My recommendation is you leave immediately for your own safety."

"No can do," said Aggie. She set the phone down and moved over to the box. "Who wants to help me here?" Steve stepped up. "Okay Steve you give me a hand. And Tom you take the phone and move out of here. Just keep the camera phone focused on what we're doing."

Steve moved up and started trying to pry the Styrofoam container up out of the larger cardboard box. He shoved his fingers down both sides for a few inches and tried to pry it with his fingernails.

"No-no, not that way. Let me show you," said Aggie. "Here's how you get a Styrofoam box out of a bigger box, Stevie baby." She picked up some thin sheets of paper. "First we carefully insert these down the sides, just to break the vacuum and make it more slippery." She handed two sheets to Steve and they slid them in as far as possible. "Okay now, we're going to hold back the top flaps of the cardboard box as we very gently roll the box upside down. Lend me a hand here. Easy. Just roll it over easy." Slowly and carefully they rolled it onto its top. "Keep those flaps out of the way...Good work. All right. Now we are going to lift this box off. You take those flaps and I'll take these. We are going to just lift it straight up. Got it?"

"Yeah, I'm good," said Steve as he squatted down and grasped two side-flaps of the box with his powerful hands.

"Ready, here we go, three, two, one, lift. Easy now, stay with me," said Aggie.

Gradually, ever so carefully, they lifted the outer box straight up and over the inner box. They stood up, together. "Don't bump it, Steve."

"We did it! Yeah! I'll just set this box over here," said Steve.

"Check inside and see if there is anything else in the box."

Steve felt around inside. "Just these papers."

"Oh good. Take those over to Tom and let him send pictures up to Cynthia and Sky." Aggie waited, catching her breath. It wasn't so much the weight of the box. It was actually quite light. But, the effort of holding herself perfectly still and her hands quiet, took its toll. By the time Steve returned she was ready to go.

"What now?" said Steve.

"Well this looks like a rather ordinary Styrofoam box. We've got it upside down, so I think we can slice the bottom away from the top with your jackknife."

Steve reached in his pocket again and opened his knife. He smiled as he held it out for her. Aggie took it and squatted down by the box. "Looks like they've sealed it up with ordinary tape. Let's just cut the tape through rather than trying to strip it off, don't you think? Less movement."

"Yeah, I can hold the thing still while you cut, okay?"

Aggie nodded and began slitting the tape. "I'm glad this knife is nice and sharp." It seemed to cut through the tape easily.

"Done," she said. "Now, we need to pry the bottom away from the top and lift it straight off without bumping the contents."

"Got it," said Steve. Aggie pried her side open about an eighth of an inch and then handed the knife to Steve. "I'll hold it still while you loosen your side up a bit."

They handed the knife back and forth a few times until they had the bottom loose. By now Steve knew what they were doing. No more words were needed. They stood up, grasped the bottom of the box, nodded, lifted it straight up and off.

Gleaming before them, resting alone on its own cover, looking small and deadly, sat the nuclear warhead.

"Don't breathe," said Aggie, as she took a careful step back.

Steve did the same. "Well, what do you know! Look at that!" he whispered.

"Wow," said Tom as he tiptoed forward, phone aimed at the device.

Chapter 15

DL Threatens President Bigelow

DL couldn't resist. He had to rub it in to that arrogant s-o-b, Bigelow. "Put me through to the president," he chortled.

The situation room fell silent as the speaker crackled on. "Sir, we have a call for you from Dear Leader of the Homeland Peninsula. Shall I activate the translator software?"

"No, just tell him I'm tied up right now and will call back."

A few seconds later the aide relayed a second message, "Dear Leader insists that you speak with him now, sir."

"Can you take a message?" asked Bigelow.

Once more the situation room waited silently.

"Sir, the message is strange."

"Okay, what is it," said Bigelow with some irritation at the repeated interruptions. "We're busy here."

"He says you have thirty minutes until the lights go out."

"What?"

"You have thirty minutes until the lights go out, sir. That's all he said and then he tittered and hung up."

"That man is crazier than a fruitcake," said Bigelow.

"Yes, sir. Thank you, sir."

Space Force Two Examines Satellite

The Space Force Two hummingbird woke up and zipped over to Western Nevada. Its pilot, Lt. Flash Gordon prepared to coordinate with the hostile satellite's orbit. "Space Force Two in position, ready for intercept," Flash spoke into his communicator.

"Roger, Space Force Two," said Colonel Sky Eastman. "We have the exact orbit now. Will paste that in for you."

Flash watched as the computer accepted the data. "Got it," he said entering the data and programming the search function. "Locked on target, sir," he reported.

"Very good," Sky responded.

"Approaching target."

"Roger."

"Will commence surveillance."

Flash made the necessary slight adjustments in his position and commenced circling the satellite, photographing a picture documentary as he proceeded.

"Those pictures are coming in loud and clear," said Sky. "They seem to compare with the pictures I have of the device we captured on Honduras Island."

Sky ran the two pictures side by side on his computer display as well as projecting them on the larger overhead display and on those of his aides and military advisors.

"Stand by while we analyze, compare and contrast the two."

Situation Room

President Bigelow received regular updates from the field force, although there seemed to be a certain "lag-time" while information was filtered through the various agencies and gatekeepers. That is why he preferred to speak directly to his field commanders and sometimes with actual combatants. Exhibiting little patience with the bureaucracy and protocol in such matters, he frequently stepped on a few toes.

"Get me Flash Gordon," he commanded.

"Yes, sir, right away, sir."

"I'll hold," said Bigelow.

Flash was very busy at the moment as was his co-pilot. The engineer was tied up as well, but he saw the incoming call from the White House out of the corner of his eye. He had just enough time to flip a switch that turned on the video feed—no time to talk.

Suddenly the situation room console lit up with a video from space. The Security Council's stuffed-shirts' eyes popped as the small, innocent-looking satellite appeared center-screen with the curvature of the incredibly beautiful blue earth in the background.

"What th' hell is that?" breathed the vice president.

No one spoke. Instead they were hypnotized by the silent video feed, as it progressed slowly around the satellite.

"What are we seeing here? Seriously, is this some joke?" asked someone.

"Let me check," said the Chief of Staff. Unsure what to do, he tried connecting with the operator, using another line.

"Did you connect us with Flash Gordon? We seem to be getting some weird video feed," he said.

"Yes but let me check the connection." He was gone for a minute. "It shows that you are connected, sir. Didn't Commander Gordon answer?"

"No voice. Nothing but this video feed."

"Let me look at it," said the young intern who happened to be the only person still working at the console today while everyone was out. "Oh, that's a video taken from space, sir. Evidently Lt. Gordon plugged you directly into their live feed for some reason. It's just some pictures they're taking of a mini-satellite." He studied the video. "Interesting," he remarked, "the way they are maneuvering around the satellite is amazing. Wow! How are they doing that? They must be too busy to talk right now."

"Do you have any idea why we are seeing this?" inquired the Chief of Staff.

"Well, I've read some things about it, sir," he replied modestly. "Must be our Space Force has special interest in this one particular object out of the eight thousand orbiting around the earth."

"Really? That many? I had no idea."

The chatty intern was trying to be helpful. "Well, a lot of it is just debris, you know. Gets to be a lot of junk up there. I've never seen them take pictures like this, though, have you? That can't be an airplane, not that high up. No airplane I know can do that. And it certainly can't be another satellite. Wow, that is so cool." The intern realized he was witnessing something extraordinary.

DL and Lee Activate

"All right, Dear Leader, it's time to push the button," said General Lee, hoping DL would finally get with the program.

"Yes, General I'm ready, but first I'll make one final warning for that dastardly Gerard Bigelow."

"Sir, may I say, please reconsider contacting the president?"

"Why? I've waited years for this."

"You must activate the bomb now, DL. NOW!" Lee was more than alarmed, he was frantic.

"Oh all right. Here we go!" With a wicked smile, Dear Leader pushed the button.

General Lee was ready to flip the switch on the real activator, but he had to be sure. "Did you push it?"

"Yes, dammit! To hell with the evil Americans!" DL laughed hysterically.

Lee grasped his activator with shaking hands. He had to use both hands to force them onto the ignition switch. His finger refused to stop trembling. Tears blurred his vision. He snuffed and wiped his eyes. It was now or never. If DL ever discovered his deception the punishment would be a slow and agonizing death only administered after witnessing some horrible things applied to his family members.

General Lee scrunched his eyes closed, breathed a hasty one-sentence prayer and pushed the switch to the on position. It was over. The bomb would be spreading destruction to much of North America. The Nuclear Winter was underway.

Would the blast go off as planned or might Kim Jung Lo's last remaining interrupt code actually work? It could be days before General Lee learned the truth. He ordered his fleet to full steam ahead. They would reach North America soon.

Flash and the Colonel

"What do you think, Colonel? Shall I capture this thing?" asked Lt. Flash Gordon, pilot of Space Force Two, the hummingbird space-ship.

"After comparing the two, there's no question. We need to disarm this baby," Sky answered. "Later we may need to alter its orbit or even capture it. But, first, let's examine it internally with the octobots. It's important to proceed cautiously, in case the thing is booby-trapped. We don't want it blowing up in your face."

"Roger, activating exam procedure," said Flash. An arm reached out from the hummingbird spaceship and launched two octobots into space on a trajectory to intercept and attach to the hostile satellite. The octobot robots were made of a soft non-metallic material that would not set off a metal detector, nor would it make any impact as it landed as gently as a tiny moth's wing. The movement of the octobot was controlled from the spaceship, as was each one of its dainty arms.

The octobots, so named because of their eight arms, landed delicately and immediately began sending back data as they minced over the surface like tiny spiders.

Space Command Headquarters' personnel watched the octobot robot-devices, enthralled to see them performing in space exactly as designed. A stream of data flowed into the mainstream computer for analysis.

Back at the White House, Tracy Youngman, the intern, was thrilled to be privy to the octobots in action. He had read about them in the National Geographic magazine in his school library. Mr. Ellis, the school librarian had been helpful in assisting Tracy to prepare for a career in robotic science.

On the other hand, inside the situation room, less-informed skeptics scoffed. "What is this, Bigelow, some kind of joke?" asked one. "I don't appreciate being played for a fool."

"Pink spiders, sure, and elephants have wings," laughed another person who had a better sense of humor.

President Bigelow remained calm, confident now that his team was on the job, defending the country. He only needed to watch and wait. He quieted the naysayers with one wave of his hand and a few words, "Quiet, you fools! Pay attention. You are witnesses to history."

Situation Room

Bigelow's infamous need to be in-control persisted in making him increasingly frustrated. "Get me Colonel Sky Eastman on the phone," he demanded without saying please or thank you.

Tracy, the young intern, wasted no time placing the call.

"Space Force Headquarters, how may I direct your call?"

"Colonel Eastman, please, the White House is calling."

"I'll connect you with his office."

"Thank you," said Tracy.

"Space Force Command. How may I help you?"

"President Bigelow calling for Colonel Eastman."

"The Colonel is unavailable, sir, I'm so sorry. There is an emergency requiring his attention."

The young intern was at a loss as to what to do. No one ever turns down the president. "Um, is the second in command available?"

"No, sir."

"Is anyone there?"

"Perhaps I can help you."

"Yes, please. Do you know what is going on?"

"Well, I'm sure the Commander in Chief knows that we are dealing with a possible attack on the United States. Right?"

"Oh," In truth Tracy had no idea what this was about. He had no top-secret clearance.

"You can tell President Bigelow that the incoming ballistic missile launched a mini-satellite which orbited over the US once and is now in its second orbit. Commander Eastman is talking directly with the Space Transporter Pilot Lt. Flash Gordon who is circling the hostile satellite taking pictures and comparing them with the sister one we captured in Honduras."

"Would that be the pictures that Lt. Gordon is beaming into the situation room?"

"Very likely."

"Thank you very much. Have Colonel Eastman call the president just as soon as he can, and please let us know if there are any developments."

"I will certainly do that."

"Thank you very much. You have been helpful. And your name?"

"Just call me Betty."

"Betty. Got it."

The intern considered exactly how to relay this news.

Honduras Island USA Team

Kelly was successful in breaking into the terrorists' computer. She quickly translated letter after letter of instructions detailing how to set up the launch vehicle, attach and program the satellite and payload. More letters contained information on the trajectory of the ballistic missile, over the United States. So far, she hadn't figured out why the Homeland was doing this. What was the purpose? Did they have some objective or were they just doing raw research? Had the terrorists been told the truth, or were they blindly following orders? Sharon relayed this information to Cynthia, who was in touch with Sky, and to Steve, who passed it on to the CIA.

Octobot

Data flowed from the octobots at an alarming rate keeping Space Headquarters scientists very busy deciphering it. At last the octobots paused awaiting further orders.

A frightening pattern emerged from the data. Hidden under the innocent looking cover was a highly explosive payload. Furthermore, the onboard triggering sequence had started, proceeding up to a point where it seemed to have hit a pause code. How long would the pause last? Would it come to a full stop, or start up again? The answers must be found and quickly.

If it was possible for a very busy and focused team of scientists to go any faster, that is exactly what happened. Space Headquarters hit overdrive and then beyond.

Sky's orders to his team went out rapid-fire.

Analyze explosives as to:
Type
Strength
Duration
Coverage
Damage estimate

206

Analyze triggering sequence.
Suggest a "stop" procedure.
Suggest and analyze a disarming procedure, as to:
Dangers
Odds of success
Analyze Capture vs. Diversionary tactics:
Pros
Cons
The answers started coming in as his team focused all their energies and intellects.
Explosive type-Nuclear
Strength-10^{24} erg
Duration-Twenty years to infinity
Coverage-North America
Damage estimate-North American power grid destroyed. Technical devices and computers meltdown. Nuclear winter.
Triggering sequence by remote control of unknown origin.
Stop by rewriting code.
Danger-Could restart firing sequence.
Possible Success unknown
Capture-dangerous
Suggest diversion over Pacific Ocean.
Conclusion: We suggest diverting satellite over remote southern Pacific Ocean or Antarctica and reprograming it to disarm.

Chapter 16

Commander in Chief

It was time to talk to the Commander in Chief. Sky opened a video conference call to the situation room. His seriously concerned face appeared on the overhead. "President Bigelow, sir, may I have a word?"

"Go ahead, Eastman. Let's hear your report," boomed Bigelow.

In the room, everyone else alerted to attention, leaning forward in their seats. This was it.

"Yes, sir. Our spaceship has intercepted a hostile satellite. You have seen the video report at the same time it came in here, I presume."

"Yes, we have, Eastman. Amazing work by our Space Force."

"Thank you, sir. That was excellent teamwork by our astronauts commanded by Lt. Flash Gordon. The hummingbird space-ship is the first of its kind, sir, thanks to you."

"So that's a hostile satellite. Just as I thought. What were those little pink spiders running around on it?" asked Bigelow.

"Those are robots, sir. They analyzed the payload and sent the results into our mainland computers. We've been interpreting those findings for the last few minutes."

"Brilliant, Colonel Eastman. I'm impressed," said Bigelow.

"Well, the results are alarming, sir, and require your decision. I'm afraid our next steps carry some risk."

"What are you saying, Eastman?"

"The satellite carries a new form of small tactical nuclear explosive which, if set off in the atmosphere, can result in an Electro-Magnetic Pulse, known as an EMP."

"I see," said Bigelow gravely, somewhat familiar with the term.

"The EMP could and probably would completely knock out the North-American power grid and destroy all computers and technical equipment in its path. Even though the bomb is now over western Nevada, we estimate that the affected area would cover most of the United States and Canada. The devastation would be enormous, as well as..."

"My God!"

There was a collective gasp in the room as the enormity of the situation sank in.

"We have devised a plan," said Sky, "which does come with risk. But there are choices."

"I see," Bigelow sighed. His body drooped for a couple seconds, and then he sat upright and squared his shoulders. "All right," he said, "tell me."

Colonel Sky Eastman drew a deep breath and spoke, "We believe the triggering sequence was started at some remote location. It went about half-way through the countdown and then paused. We don't know how or why that happened, and so we do not know whether it is programed to start up again, or whether the remote controller might set it off. The satellite will orbit the earth again about every ninety minutes, and so the enemy has figured out where they want it to explode, no doubt. Our people are writing a new sequence of instructions for the onboard computers, which, we hope will disarm the device.

209

"Option number one: we can attempt to enter those into the computer's triggering sequence, using the octobots. If you order that, we suggest diverting the satellite over a remote area of the Southern Ocean or over Antarctica before we attempt this procedure, in case it explodes. Or, option two, we can try to capture the device. Finally, option three, we can simply watch and follow it until it decays of its own accord and falls to earth."

For once Bigelow was quiet. All eyes were on him. The very fate of the world rested on this decision.

"Tell me again, Sky, why we don't just capture this thing and destroy it?"

"Well, we can try that, sir, but what if we cause it to go off? Where would we take it? Do we have a safe place to set off an atomic explosion, even a small one?"

Bigelow rested his head on his hand.

"And so, are you saying we can divert its path?" asked the solemn president.

"The device is on a LEO, low earth orbit, which will take it around the globe in a north-south configuration every ninety minutes. Also, since the earth is spinning at the same time, each orbit will take the satellite further to the west by one and one-half time zones. The next orbit will be out over the Pacific, just west of our coast. The problem is that, unless something is done, there is enough drag on the satellite to gradually slow it down and into the atmosphere where it could disintegrate and fall to earth. Or, it could explode."

"Do we know how long it will stay up?"

"We are working on that, sir, but I don't think we can predict within more than a day or two. In the course of two days it will have covered every point on the earth about three times."

"Meaning it could come down anywhere."

"That's it, sir."

"So if it is going to come down, it's better to control that, right?"

"You got it."

"Wouldn't it be best to bring it down somewhere out in the middle of the ocean?"

"Not entirely."

"Why is that?"

"First we have to find a fairly wide area completely free of ships, or islands, in case the device explodes. And there is the contamination that will result in that area of the world, and the radiation drift which can encircle the globe. Also, if we want to have access to the device for study, we need to capture it."

"Hmm," said Bigelow, "I see what you're saying." He wrinkled his brow. "But, didn't I learn there is a similar device that we already have in custody, on an island, somewhere or other?"

"Yes, sir. We captured a terrorist group on an island off the coast of Honduras. We have the group in custody now. Also, we stopped them from setting off their ballistic missile. We believe the device is similar. However, we only have the report of one CIA agent on the ground at that location."

"Surely he didn't capture these people all by himself, Eastman."

"No, sir. There were some American civilians with him at the time. However, he kept them out of danger and used a military force from Guantanamo Bay to wrap up the operation."

Meanwhile, Sky knew the hostile satellite was orbiting away from the continent. He was worried that this was taking too long. Nevertheless, he had to serve the president, helping Bigelow make his decision.

"Civilians? What civilians?" Bigelow questioned.

"It was a private security group out of Senator McBride's office, sir, plus a cop from New Mexico and his bride."

"Good grief, Eastman, I might have known McBride was mixed up in this somehow. That man has rescued us before."

"Yes, sir."

"And your wife is part of this, I'll wager."

"Yes, sir. It was her group that tipped us off."

"Humph...well then..." Bigelow cleared his throat. "Why can't we just dump this weapon into the ocean and study the one in Honduras instead?"

"That is an option, sir."

Bigelow paused. His mind was putting two and two together.

"Look here, Eastman, you must have some idea who is behind all this."

"Well, so far as the Honduras missile is concerned, we know who is behind it. But, we have no real evidence that the same country set off the present satellite. That is only supposition due to the similarities."

"Didn't you determine the two satellites look alike?"

"Yes, sir."

"Well then, Eastman, use your common sense. It must be the same outfit." Bigelow decided. "Who was it?"

"Sir, Senator McBride's sister has broken into the Honduran terrorist's computer system. She has translated many documents, all showing that General Lee of the Homeland Peninsula was directing the operation. The terrorists are being interrogated at Guantanamo, but, so far, they're saying nothing."

"Legal interrogations take time."

"Indeed they do, and we are losing time, sir," Sky dared to remind the president.

"I'll get back to you," said President Bigelow, abruptly, to Sky's astonishment. Didn't he realize the urgency? The connection went dead.

Suddenly bursting with energy and resolve, President Gerard Bigelow ordered, "Get me Dear Leader on the phone, stat!"

Shoe is on the Other Foot.

"Got your message, DL you crafty son-of-a-bitch!"

DL blanched. Who is this? Bigelow? How did he get through? "W-who's c-calling?" he stammered.

"This is the President of the United States, you bastard!"

DL gulped. The lights should be out in North America by now. Maybe this was a hoax. Yes, that had to be it.

"See here, whoever you are, the penalty for treason against the regime is swift and painful," DL blustered. "I recommend you get your affairs in order, quickly. My men are on their way."

"Ha! Not so fast, DL. For your information, your little bomb is on *its* way back to you. The lights are still burning brightly here in North America. By my calculations you have exactly thirty-five minutes until your puny little satellite orbits back over your side of the world. And the United States Space Force has plans to put it in a permanent orbit directly over your office. Do you hear me, DL? Directly over the Homeland Peninsula!"

"P-preposterous!" DL protested. "I pulled the trigger myself."

"Is that so? Well, maybe you'd better check with your General Lee and find out why that didn't work. Also, that other little present you had for us on that island off the coast of Honduras. Well, we're looking at that right now, and thinking maybe we'll send that one up over your head, too, just as a back up."

"I don't believe you." DL puffed.

Bigelow paused to read a note that his Chief of Staff handed him. Laughing out loud, "Oh DL, by the way, your General Rhee Su-jin and her team are in custody at our special hotel on Guantanamo Bay. She's enjoying her freedom and is only too happy to cooperate with us."

Dear Leader seemed to be wilting.

"Oh, one more thing, DL, speaking of nuclear, our nuclear submarines, the Nevada and the Nebraska, are tracking your pathetic fleet at this very moment. I recommend you recall your army if you don't want it buried at sea. We'll be more than happy to escort your troop ships back to the Homeland."

DL sighed and shriveled back into his chair, the telephone dangling from his hand.

"I'll be leaving you now, DL," said the president. "Have a nice day."

Epilogue

Homeland Army is Recalled

An official welcoming committee was waiting for General Lee when he walked down the gangplank at their home port. There would be no brass band, no medal ceremony. Instead he was led off in handcuffs, ushered into a black limo with darkened windows and slowly driven far away. He was never seen again.

The chief scientist, Kim Jung Lo made it safely back to his farm, where he lived in obscurity, unaware his single line of code had stopped the activation sequence and succeeded in saving the world.

Senate Offices

Back in DC, Aggie and Sharon had spent the first lunch hour filling-in the guys and gals on a few of the less exciting details. Much of what they knew was classified, and truthfully, even they did not know all that happened.

"Tell them about the good-looking guy you met, Sharon," Agatha teased.

"Nonsense," Sharon denied. "I have no idea who you mean."

Just before closing time, a dozen roses arrived on Sharon's desk. "Now who could have sent these," she mused. All the ladies looked on with curiosity as Sharon reached for the card. It read: 'Dinner? Meet me

215

downstairs. Signed Steve Somebody.' She allowed herself a secret smile and slipped the card into her bosom.

"Who's it from?" chorused the nosey crowd.

Sharon merely bent over to fetch her purse from the bottom drawer. "Oh just somebody," she tossed over her shoulder as she headed for the door. "Have a nice evening, everyone."

Arlington, Virginia

After finishing the mission, Sky Eastman was summoned to Washington by the Commander-in-Chief. Arriving at her apartment, he dropped his bags in the foyer and turned to greet Cynthia. "Hi Darling, am I late?" he asked, wrapping her in his arms.

"You're good," she answered, pressing her body to his dress uniform. "I'm almost ready to go and you look perfect for dinner at the White House." She wolf-whistled while leaning back for a good look. "Whoa, what's this?" She touched one manicured fingernail to a shiny new star on her husband's jacket. "Is this what I think it is?"

Sky grinned. "Ah well, no big deal. It's only one star," he replied modestly.

"Well, you deserve at least two, maybe even three." Cynthia pretended to pout.

"Not me," said Sky. "Credit goes to my computer experts who wrote the successful code. Mostly I just sat back and watched Flash Gordon and his crew, Jim Hughes and the others."

"How so?" asked Cynthia eager to learn what happened.

"Well, it took some nerve to carry out Bigelow's orders."

"Oh?" Cynthia raised an eyebrow. "Orders?"

"Let's just say the United States Space Force successfully disarmed two warheads and recovered one suspicious satellite."

"Wow." Cynthia declared. "That's amazing!"

"Yup," Sky agreed.

Glancing at her watch, she noted, "Well, sweetheart, the President's limo will be here to pick us up in about ten minutes."

"I'll be ready in five," said Sky, as he headed down the hall.

Tom and Kelly Choose Wisely

Sgt. Tom Turbulo and his new wife Kelly (McBride) Turbulo accepted a congratulatory phone call from President Gerard Bigelow. In the end, they politely declined an invitation to join the other members of the Honduras USA team with Agent Steve Spalding, Senator Mike McBride and his wife Juliette, and Brigadier General Sky Eastman and his wife Cynthia Patterson, for a special dinner hosted by the Bigelows at the White House.

Instead they opted for an extra three weeks' honeymoon and a lift on Air Force One to a seven-hundred-dollar a day ultra-safe, ultra-secret posh island resort of their choice, courtesy of the President of the good old US of A.

The End

Dear Reader;

If you enjoyed reading this book we know you will want to thank the author. Please take a moment to do so by leaving a review on the web page where you ordered the book. We promise she reads all of them and usually responds.

More wonderfully entertaining books just like this one, by Dorothy May, can be found online, at your favorite library or bookstore, and at MercerPublications.com website.

email: info@MercerPublications.com

Our gift to you, just for reading this book:
For a discount coupon on your next Mercer
Publications entertaining novel, go here:
www.MercerPublications.com

See a partial list of available titles:

BOOKS FROM MERCER PUBLICATIONS
The Complete *How to For You* Series:

By Dorothy May Mercer
Links to these books can be found at
www.mercerpublications.com

The complete "How to For You" series of E-booklets for improving writers.

How to Write Fiction
How to Write Great Dialog
How to Design & Format Your Paragraphs
How to Write Sentences and Paragraphs in Your Novel
How to Fix Errors in Your Document
How to Format Your Book, for Publishing
How to Sell Your eBook Using Free Days
How to Add an Interactive Table of Contents
How to Install a Link in Your Document
How to Edit a Book, *With a Friend*
How to Prepare Your Book for eBooks
How to Use Your Book for Free Ads
How to Design an eBook Cover
How to Install eBook Cover on Print Books, and Vice Versa
How to Market Your Book
How to Register ISBNs & Copyrights
How to Get an Audible Version, of Your Book
How to Self Publish, *Your Book*
How to Create a Picture Book

The complete "How to For You" PRINTED books for improving writers.

How to Write Fiction
How to Write Great Dialog
How to Edit a Book, *With a Friend*
How to Format Your Book, *for Publishing*
How to Get an Audible Version, *of Your Book*
How to Market Your Book
How to Register ISBNs & Copyrights
How to Self Publish, *Your Book*
How to Create a Picture Book

More Entertaining Books from Mercer Publications, Inc.

In business since 1993

by Dorothy May Mercer and other distinguished authors:

McBride novels available—eBook, Print, Audible, English and
    Spanish formats.

Starring Det. Lt. Mike McBride and his cop buddies, as they solve
crimes and fall in love, one at a time.

The McBride Series of Action Novels, Starring Det. Lt. Michael J.
McBride Jr., a good cop, his buddies and a gorgeous redhead,
available in English and Spanish, eBook, print and Audible editions

"Car oo6 Responding"   "Unidad oo6 Respondiendo"
"The Cocaine Chase"   "La Casa di la Cocaina"
"The Golden Coin"   "El Immigrant e la Monada Dorada"
"The Cartel Wars"   "La Guerras Cartel"
"The Gang Bust"   "La Pandilla Busto"

The Washington McBride Series

Starring Senator Mike McBride, available in eBook, Print and Audible versions.

- "The Fairfax Fix"
- "The Arlington Alias"
- "The Savage Surrogate"

Starring wife Juliette McBride, investigative reporter and Lady Dog, famous Seal-trained tracking and service dog.

The New McBride Suspense (with Romance) Series

Fran and Max, *the Bungalow*

Hidden away in a specially fortified bungalow, Fran awaits the birth of her baby. Will the syndicate find her? A handsome FBI agent complicates her life.

Mary Beth and Sammy, *Rolling Thunder*
Cynthia and Dan, *Cyber War*

Nate, *The Search*

A father's search for his long-lost daughter, leads to an evil terrorist's plot to take down US airplanes with amazing new hi-technology. As usual, there is a bit of romance, as well.

E M P Honeymoon, *Kelly & Tom*

An innocent new bride stumbles unto a terrorists' plot aiming to blow up the USA with hi-tech but untested warheads. It takes the US Space Force, a handsome CIA agent and the McBride ladies to come to the rescue.

GO HERE TO ORDER THIS AND PREVIEW OTHER EXCITING AND ENTERTAINING NOVELS:

- http://www.MercerPublications.com

Historical books by Dorothy May Mercer:

"Leon and Esther,"

a Christian love story, set in the 1920s, in rural Michigan.
Mature readers, please..

"Stories I Haven't Told," autobiography

From barefoot farm girl to multi-millionaire CEO in America.

Travel Books with Colored Photographs:

<u>By Dorothy May Mercer and Photographer Dave Mercer</u>
  "<u>Alaska and Back</u>" a travel journal.

"<u>Hawaii and Back</u>" Vol. 1 <u>Kauai</u>, With Dave and Dorothy

"Hawaii and Back" Vol. 2 Maui, With Dave and Dorothy

"Hawaii and Back" Vol. 3 Oahu, With Dave and Dorothy

Africa and Back,
With Dave and
Dorothy

Niagara and
Back"
With Dave and
Dorothy

More Books edited and published by Mercer Publications & Ministries, Inc.:

~ *Recommended* ~

- "He Called Her Hat," That Tough Little Lady, by Myron C. McDonald

- "Notes from John," [and]
- "Ascension Teachings with Archangel Michael," [and]
- "With Love from Diana, Queen of Hearts," [and]
- "Mary Magdalene Speaks, all by Marcia McMahon

- "Let's Talk" a Black & White Dialog in US & UK, by Anonymous

- "Short & Fun Stories" Vol 1 and Vol 2, by Various Authors

- "Without from Within," by Ron Shaw
- "The Yellow Bus Boys," by Ron Shaw

- "Remember How Much I Love You," by Dale L. Williams M.D.

- "The Inheritance from Hell," by R.D. Margot

- "Stormy Affair" [and]
- "Obsession" [and]
- "Sensual Bond" Vol. 1-5,all by Netty Ejike

- "Gems" by Nancy Calumet

- "Him and Her" by Gerald Kinsey

Characters

Ch 1

Kelly and Tom Turbulo. Newlyweds. Tom is a cop in the CCPD.

Mike McBride. Senator from New Mexico

Cynthia Patterson-, his secretary, bodyguard and private investigator

Dr. Joseph "Jo" Rench, CIA Assistant Director

Burly guy, tackles Kelly

Taxi driver, speaks Spanish and English

First mate on the dive boat

Ch. 2

Professor Dr. Peter Kinney, Nobel Award winning nuclear physicist, expert in the power grid

Henry Haven-Harbinger, talk show host

Ch. 3

Colonel Rhee Su-jin alias Sue Lynn Reese

Dear Leader, a.k.a. DL

Aide to Leader

Beach Attendant

Steve Spalding, alias tall bronze man, CIA Special agent

Ch 4

Clerk at scooter rental place

Barefoot waiter

Lt. Sam Mulholland (non-speaking)

Ch 6

Two men terrorists who work for Su-jin

Ch 9

Mitch Mucurie, photo analysist CIA

Ch 11

Major/Colonel Sky Eastman, Cynthia's husband.

Jake, Sky's aide.

James Petrie, Pentagon photo-analyst
Ch 12
US President Gerard Bigelow
Mrs. Beth Terry, his personal secretary.
Ch 13
Major Jim Hughes, Pilot Space Force One
Lieutenant Andrew "Flash" Gordon, space-transporter
pilot of Space Force Two
Betty, Space Headquarters worker
Tracey Youngman, White House intern

Sources and Research

Every day, the United States Space Command (USSPACECOM) monitors more than 8,000 man-made objects orbiting the Earth. The command supports manned space flight with data from a worldwide Satellite Surveillance Network (SSN) of 17 radar and optical sensors. These provide extensive worldwide but not complete global coverage. The sensors are able to monitor an object as small as a baseball in Low Earth Orbit approximately 100 to 600 miles from Earth and as small as a volleyball in Geosynchronous Orbit, approximately 22,300 miles.

USSPACECOM supports manned missions like our Space Shuttle and the Russian Mir Space Station, comparing the orbits of satellites and debris with the orbit of manned spacecraft 36 hours into the future. This support is also provided for satellites which are to be deployed by the shuttle to conduct an experiment or research and then retrieved by the shuttle. Prior to the shuttle launch and any on-orbit maneuvers, the National Aeronautics and Space Administration (NASA) will inform the Space Control Center in Cheyenne Mountain who will in turn check for any possible close approaches in the new flight path 36 hours out. If the orbit of any satellite or debris comes close to the predicted orbit of the Shuttle or Mir, USSPACECOM immediately notifies NASA. If the close approach is to the shuttle, NASA decides whether to maneuver the shuttle to avoid the satellite or debris. If the Mir is involved, NASA will forward the information provided by USSPACECOM to the Russian Space Agency and the Russians will determine the appropriate action.

Also closely supported by USSPACECOM are shuttle missions involving space rendezvous

The fighter radar needs higher resolution and processing power to handle that range. It is usually looking for fast moving distant small targets in the expanse of sky. When it finds something it needs to determine the targets bearing and direction of travel, compute an intercept, and track it, as well as many others at the same time, and define the overall threat analysis to the pilot.

Low Earth orbit

From Wikipedia, the free encyclopedia
Comparison
of geostationary, GPS, GLONASS, Galileo, Compass (MEO), International Space Station, Hubble Space Telescope and constellation orbits, with the Van Allen radiation belts and the Earth to scale.[a] The Moon's orbit is around 9 times larger than geostationary orbit.[b] (In the SVG file, hover over an orbit or its label to highlight it; click to load its article.)

A low Earth orbit (LEO) is defined by Space-Track.org as an Earth-centered orbit with at least 11.25 periods per day (an orbital period of 128 minutes or less) and an eccentricity less than 0.25.[1] Most of the manmade objects in space are in LEO orbits.[2] A histogram of the mean motion of the cataloged objects shows that the number of objects drops significantly beyond 11.25.[3]

There is a large variety of other sources[4][5][6] that define LEO in terms of altitude. The altitude of an object in a elliptic orbit can vary significantly along the orbit. Even for circular orbits, the altitude above ground can vary by as much as 30 km (19 mi) (especially for polar orbits) due to the oblateness of Earth's spheroid figure and local topography. While definitions in terms of altitude are inherently ambiguous, most of them fall within the range specified by an orbit period of 128 minutes because, according to Kepler's third law, this corresponds to a semi-major axis of 8,413 km (5,228 mi). For circular orbits, this in turn corresponds to an altitude of 2,042 km (1,269 mi) above the mean radius of Earth, which is consistent with some of the upper limits in the LEO definitions in terms of altitude.

The LEO region is defined by some sources as the region in space that LEO orbits occupy.[7][8][9][10] Some highly elliptical orbits may pass through the LEO region near their lowest altitude (or perigee) but are not in an LEO Orbit because their highest altitude (or apogee) exceeds 2,000 km (1,200 mi). Sub-orbital objects can also reach the LEO region but are not in an LEO orbit because they re-enter the atmosphere. The distinction between LEO orbits and the LEO region is especially important for analysis of possible collisions between objects which may not themselves be in LEO but could collide with satellites or debris in LEO orbits.

The International Space Station conducts operations in LEO. All crewed space stations to date, as well as the majority of satellites, have been in LEO. The altitude record for human spaceflights in LEO was Gemini 11 with an apogee of 1,374.1 km (853.8 mi). Apollo 8 was the first mission to carry humans beyond LEO on Dec 21-27, 1968. The Apollo program continued during the four-year period spanning 1968 through 1972 with 24 astronauts who flew lunar flights but since then there have been no human spaceflights beyond LEO.

Orbital characteristics
The mean orbital velocity needed to maintain a stable low Earth orbit is about 7.8 km/s but reduces with

increased orbital altitude. Calculated for circular orbit of 200 km it is 7.79 km/s and for 1500 km it is 7.12 km/s.[11] The delta-v needed to achieve low Earth orbit starts around 9.4 km/s. Atmospheric and gravity drag associated with launch typically adds 1.3-1.8 km/s to the launch vehicle delta-v required to reach normal LEO orbital velocity of around 7.8 km/s (28,080 km/h).[12]

People and objects in orbit experience weightlessness because they are in free fall even though the Earth's pull due to gravity in LEO is not much less than on the surface of the Earth.
Objects in LEO encounter atmospheric drag from gases in the thermosphere (approximately 80-500 km above the surface) or exosphere(approximately 500 km and up), depending on orbit height. Due to atmospheric drag, satellites do not usually orbit below 300 km. Objects in LEO orbit Earth between the denser part of the atmosphere and below the inner Van Allen radiation belt.
Equatorial low Earth orbits (ELEO) are a subset of LEO. These orbits, with low inclination to the Equator, allow rapid revisit times and have the lowest delta-v requirement (i.e., fuel spent) of any orbit. Orbits with a high inclination angle to the equator are usually called polar orbits.

Wikipedia/Solar flare "On July 23, 2012, a massive, and potentially damaging, solar superstorm (solar flare, coronal mass ejection, solar EMP) barely missed

Earth, according to NASA.[5][6] According to NASA, there may be as much as a 12% chance of a similar event occurring between 2012 and 2022,[5] although because this particular figure was based on an extreme extrapolation of the calculated frequency of future storms, the actual probability of this is quite uncertain."

Source Wikipedia superflare

Even for much smaller superflares, at the lower end of the Kepler range, the effects would be serious. In 1859 the Carrington event caused failures in the telegraph system in Europe and North America. Possible consequences today would include:

Damage to or loss of all artificial satellites

Airline passengers on trans-polar flights would receive high radiation doses from the energetic particles (as would any astronauts or the crew of the International Space Station)

Significant depletion of the ozone layer with increased risk of cataracts, sunburn and skin cancer, as well as damage to growing plants. The recovery time would be of the order of months to years. In the strongest cases there would be severe damage to the biosphere, especially to primary photosynthesis in the oceans

Failure of the electricity distribution system (as in the March 1989 geomagnetic storm), possibly with damage to transformers and switching equipment

Loss of power to the cooling systems of spent fuel rods stored at nuclear power stations

Loss of most radio communication because of increased ionization in the atmosphere

It is evident that superflares often repeat rather than occurring as isolated events. The NO and other odd nitrogens created by flare particles catalyze the destruction of ozone without being consumed themselves and have a long lifetime in the stratosphere. Flares at a frequency of one a year or even less would have a cumulative effect; the destruction of the ozone layer could be permanent and lead to at least a low-level extinction event.

Superflares have also been suggested as a solution to the Faint young Sun paradox.[27]

Can superflares occur on the Sun?

Since superflares can occur on stars apparently equivalent in every way to the Sun, it is natural to ask if they can occur on the Sun itself. An estimate based on the original Kepler photometric studies suggested a frequency on solar-type stars (early G-type and rotation period more than 10 days) of once every 800 years for an energy of 1034 erg and every 5000 years at 1035 erg.[3] One-minute sampling provided statistics for less energetic flares and gave a frequency of one flare of energy 1033 erg every 5-600 years for a star rotating as slowly as the Sun; this would be rated as X100 on the solar flare scale.[5] This is based on a straightforward comparison of the number of stars studied with the number of flares observed. An extrapolation of the empirical statistics for solar flares to an energy of 1035 erg suggests a frequency of one in 10,000 years.

However, this does not match the known properties of superflare stars. Such stars are extremely rare in the Kepler data; one study showed only 279 such stars in 31,457 studied, a proportion below 1%; for older stars this fell to 0.25%.[3] Also, about half of the stars which were active showed repeating flares: one had as many as 57 events in 500 days. Concentrating on solar-type stars, the most active averaged one flare every 100 days; the frequency of superflare occurrence in the most active Sun-like stars is 1000 times larger than that of the general average for such stars. This suggests that such behavior is not present throughout a star's lifetime but is confined to episodes of extraordinary activity. This is also suggested by the clear relation between the magnetic activity of a star and its superflare activity; in particular, superflare stars are much more active (based on starspot area) than the Sun.

There is no evidence for any flare greater than the Carrington event (about 1032 erg, or 1/10,000 of the largest superflares) in the last 200 years. Although larger events from the 14C record ca. 775 AD is unambiguously identified as a solar event, its association to the flare energy is unclear, and it is unlikely to exceed 1032 erg.

National Geographic Magazine May 2017

An article: "The Softer Side of Robots," by Natasha Daly, discusses the "octobot" robot. "The octobot is the world's first completely soft, autonomous, and untethered robot. It is free of wires, batteries and any hard material—like its namesake, the octopus." The article goes on to describe the tiny octobot and its possible amazing future uses.

$9.99

Made in the USA
Monee, IL
18 January 2020